An Holistic Guide to Anatomy & Physiology

Hairdressing and Beauty Industry Authority Series – related titles

Hairdressing

Start Hairdressing: The Official Guide to Level 1
Martin Green and Leo Palladino

Hairdressing: The Foundations – The Official Guide to Level 2
Leo Palladino

Professional Hairdressing: The Official Guide to Level 3
Martin Green, Lesley Kimber and Leo Palladino

Men's Hairdressing: Traditional and Modern Barbering Maurice Lister

African-Caribbean Hairdressing Sandra Gittens

The World of Hair: A Scientific Companion Dr John Gray

Salon Management Martin Green

Essensuals, Next generation Toni & Guy: Step by Step

Mahogany Hairdressing: Steps to Cutting, Colouring and Finishing Hair
Martin Gannon and Richard Thompson

Mahogany Hairdressing: Advanced Looks
Martin Gannon and Richard Thompson

Patrick Cameron: Dressing Long Hair Patrick Cameron and Jacki Wadeson

Patrick Cameron: Dressing Long Hair Book 2 Patrick Cameron

Bridal Hair Pat Dixon and Jacki Wadeson

Trevor Sorbie: Visions in Hair Kris Sorbie and Jacki Wadeson

The Total Look: The Style Guide for Hair and Make-Up Professionals
Ian Mistlin

Art of Hair Colouring David Adams and Jacki Wadeson

Beauty Therapy

Beauty Therapy: The Foundations – The Official Guide to Level 2
Lorraine Nordmann

Professional Beauty Therapy: The Official Guide to Level 3
Lorraine Nordmann, Lorraine Appleyard and Pamela Linforth

Aromatherapy for the Beauty Therapist Valerie Ann Worwood

An Holistic Guide to Anatomy & Physiology Tina Parsons

Indian Head Massage Muriel Burnham-Airey and Adele O'Keefe

The Complete Nail Technician Marian Newman

The Encyclopedia of Nails Jacqui Jefford and Anne Swain

The World of Skin Care: A Scientific Companion Dr John Gray

Safety in the Salon Elaine Almond

An Holistic Guide
to Anatomy &
Physiology

Tina Parsons

HABIA
Hairdressing And Beauty Industry Authority

THOMSON

Australia • Canada • Mexico • Singapore • Spain • United Kingdom • United States

THOMSON

An Holistic Guide to Anatomy & Physiology

Copyright © Tina Parsons 2002

The Thomson logo is a registered trademark used herein under licence.

For more information, contact Thomson Learning, High Holborn House, 50-51 Bedford Row, London WC1R 4LR or visit us on the World Wide Web at: http://www.thomsonlearning.co.uk

British Library Cataloguing-in-Publication Data
A catalogue record for this book is available from the British Library

ISBN 1-86152-976-7

First edition published 2002 by Thomson Learning
Reprinted 2003 and 2005 by Thomson Learning

Typeset by 🅣 Tek Art, Croydon, Surrey
Printed and bound in Singapore by Seng Lee Press

Contents

The genito-urinary system

The nervous system

The endocrine system

Foreword

In life we start by thinking we can do everything, but in reality we know very little. And what we do know, we rarely put into practice. As you grow older you start to take more notice of your body. It is not to do with vanity but more to do with new aches, pains and a general feeling of slowing down.

Tina Parsons has written a marvellous book, which explains many of the questions we ask ourselves about our bodies. This comprehensive guide bridges the gap between theory and practice for both clients and therapists. It is aimed at students of beauty and holistic therapies, but is so full of knowledge and fascinating facts that many an experienced educator will want it continually at their side. This book is well written and I found it so engaging that it created a desire to learn more about the human body. It was difficult to put down.

HABIA is dedicated to learning. This book advances our philosophy, for all within the hair and beauty industry. Tina is an accomplished writer, passionate about her subject. She is an example for us all to follow. Read this book and educate yourself.

Alan Goldsbro
Chief Executive Officer
HABIA

Introduction

Learning is all about being brave enough to ask questions and the inspiration for this book lies with all the students, staff and clients who have ever asked me 'why'? You know who you are!

This book has therefore been written for *students* of beauty and holistic therapies to cover the anatomy and physiology content of your courses. It provides a link between theory and practice, which is sometimes difficult to appreciate when first embarking on a course of this nature. This book has also been written to provide a back up text for those *qualified therapists* working in the industry. It will provide a useful point of reference to reinforce and enhance your knowledge. Finally, the target audience for the book includes our *clients* – those people who actively take an interest in the workings of their body and want to know more. It will make interesting reading in the reception of any salon helping to promote the provision of beauty and holistic services.

Together we are going to take a fascinating journey through the workings of the human body stopping off at each major body system to explore its specific structures, i.e. **anatomy**, and functions, i.e. **physiology**.

The major body systems include:

● The Integumentary System

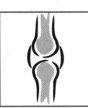

● The Skeletal System

● The Muscular System

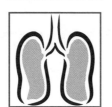

● The Respiratory System

● The Circulatory Systems

● The Digestive System

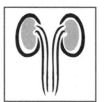

● The Genito-Urinary System

● The Nervous System

● The Endocrine System

In order to guide you through the journey of the human body, a chapter has been devoted to each of these body systems and each chapter has been subdivided into six sections including *science scene, common conditions, system sorter, holistic harmony, treatment tracker* and *knowledge review*.

- **Science Scene** includes detailed explanations of the structures and functions of each system of the body providing you with the level of underpinning theoretical knowledge required to perform beauty and holistic treatments to aid the whole of the human body.

- **Common Conditions** includes information on the types of conditions that can affect each system of the body, providing a guide that can be used to ascertain whether or not a client is **indicated** for treatment i.e. the proposed treatment will aid their condition, or **contra indicated** i.e. the proposed treatment may make their condition worse. It also provides a point of reference to the various conditions relating to each body system as well as a guide for client referrals.

- **System Sorter** includes a quick and easy reference to the ways in which all of the body systems link together. This will help to develop an awareness of the fact that body systems do not work in isolation; they need the continuing support of one another to keep in healthy working order.

Throughout the text there are references to some *Fascinating Facts* that are associated with each individual body system.

- **Holistic Harmony** includes information about the needs of the individual body systems illustrating that there is more to a person than their anatomical and physiological make up. The information is subdivided into nine different categories including *fluid*, *nutrition*, *rest*, *activity*, *air*, *age*, *colour*, *awareness* and *special care*. This section provides the vital information that can be used to support beauty and holistic treatments and it forms a basis for the after care advice given at the end of a treatment.

Tip

Providing clients with long-term treatment goals is not only good practice professionally but also commercially.

- **Treatment Tracker** includes a quick and easy reference to the ways in which all types of beauty and holistic treatments are of benefit to all of the systems of the body. The effects of treatments are progressive and cumulative so this information provides the basis for the recommendation of further treatments.

- **Knowledge Review** provides you with a chance to test your knowledge of the individual body systems with a set of short answer questions. The questions are designed to ensure that you have gained the appropriate level of understanding of the anatomical and physiological theory that underpins the wide range of beauty and holistic treatments.

Tip

A working knowledge of the human body is a necessary tool for every therapist.

Remember

Remember, that as therapists we do not diagnose medical conditions. Instead, we refer clients to their GP or specialist in the event of a suspected or known condition that is thought to be contra indicated to treatment.

Website

In addition to the information provided by the book, there is also access to an interactive website for student and lecturer use. The website will provide a back up to learning by supplying the answers to the *Knowledge Review* questions as well as assignments for each chapter. It has been designed to provide as much access to varied learning as possible in an attempt to meet the growing needs of individual learners. Lecturer support will also be provided in the form of interactive input facilitating the use of techniques for teaching, assessing and checking of the learning process.

Angel advice

Good teaching is all about inspiration, intuition and innovation and good teachers leave a part of themselves with every person they have the privilege to teach; key words, deeds and care remain with us long after the initial learning has taken place, helping us shape our future, rather like a guardian angel of learning. With this in mind, the book includes helpful tips and hints in the form of *Angel Advice* on how to build awareness and self-discovery in ourselves and our clients.

Acknowledgements

Finally, I have appreciated the opportunity that Thomson have given me to write this book and would like to dedicate it to my parents for creating the footsteps for me to follow.

Tina Parsons
2001

The cells

Learning objectives

After reading this chapter you should be able to:

- **Recognise the structure of a cell**

- **Identify the different tissue types**

- **Understand the functions of cells**

- **Be aware of the factors that affect the well-being of the cells**

- **Begin to appreciate the workings of the human body in relation to the other systems of the body.**

Before our journey through the human body can begin, we need to gain a general picture by looking at the structure and function of the minute parts that make up each of the individual systems – the **cells**.

The body is made up of millions and millions of individual *microscopic* cells which are so tiny that they are only visible through a microscope, hence the term microscopic. Cells are the building blocks with which the human body is formed. They build into **tissue**, **organs** and **glands**, **body systems** and finally the human **organism**.

Science scene

Structure of cells, tissues, systems

Cells

Cells vary in size and shape, but they have a basic structure which is common to most types of cells.

Cells are made of **protoplasm,** which is a colourless, transparent jelly-like substance consisting of approximately 70 per cent water together with organic and inorganic substances.

The structure of a cell

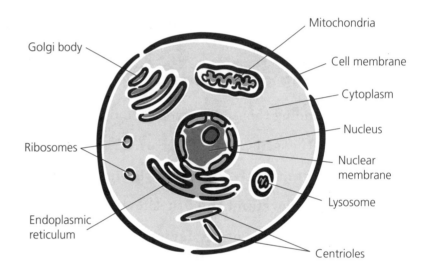

Most cells consist of three main sections: an outer layer called the **cell membrane,** an inner **nucleus** and a middle layer of a semi fluid substance called **cytoplasm.**

1. The cell membrane is made up of proteins and fats and is semi permeable i.e. it allows some substances like oxygen and carbon dioxide to pass through it.

2. The nucleus lies in the centre of a cell. It is made up of a special form of protoplasm called **nucleoplasm**. The nucleus is often referred to as the 'information centre' of a cell as it contains all the instructions for the growth, development and function of the cell in the form of **DNA** (**d**eoxyribo**n**ucleic **a**cid). DNA carries the material needed to form **chromosomes**, which carry the inherited information that is passed on by parent cells. Human cells contain 46 chromosomes, 23 from each parent. The nucleus is surrounded by a nuclear membrane separating it from the other structures within the cell.

3. The cytoplasm contains many tiny structures known as **organelles** or 'little organs' including: **mitochondria, ribosomes, golgi body, lysosomes, endoplasmic reticulum** and **centrioles.**

- Mitochondria are small spherical, rod shaped structures, which are often referred to as 'power houses' because they provide the cell with the power needed to create energy.
- Ribosomes are granular structures, which are often referred to as 'protein houses' because they provide the cell with the protein needed for its growth and repair.
- Golgi body comprises of four to eight stacked sacs, which process, sort and deliver proteins to other parts of the cell to be used as energy.
- Lysosomes are spherical structures that produce substances to break down damaged and worn out parts of the cell. Often referred to as the cell's 'disposal units'.
- Endoplasmic reticulum are a series of canals transporting the different substances around the cell.
- Centrioles are two tiny cylindrical structures that lie at right angles to one another. They are involved with the reproduction of new cells.

Cells do not operate on their own; instead, they work together in groups of similar types of cells to form **tissue**.

Tissue

There are four different types of tissue including: **epithelial**, **connective**, **muscular** and **nervous**.

Epithelial tissue

Epithelial tissue forms the linings or coverings of many organs and vessels of the body and can be subdivided into two types – **simple** and **compound epithelium.**

1. Simple epithelium is comprised of a single layer of cells and comes in four different varieties:
   - **Squamous** or **pavement** – flattened, scale-like cells arranged edge to edge in a row rather like a tiled floor. Squamous epithelium forms parts of the body that have very little wear and tear e.g. the lining of the alveoli of the lungs in the respiratory system and the linings of the heart, blood and lymph vessels in the circulatory systems.
   - **Cuboidal** – cube-shaped cells arranged in a row to form the linings of some glands. This tissue releases fluids as part of a process called **secretion** e.g. sweat from a sweat gland.

- **Columnar** – a row of tall cells, which form the linings of many of the parts of the digestive and urinary systems. Specialist cells called **goblet cells** are found amongst the columnar cells, which secrete a watery fluid called **mucus**.

- **Ciliated** – a single row of squamous, cuboidal or columnar cells containing fine hair-like projections called **cilia.** The cilia move regularly in a wave-like motion all in the same direction, which helps to move substances along them e.g. mucus, unwanted particles etc. The linings of the respiratory system and parts of the reproductive system are formed from this type of epithelial tissue.

Simple epithelium

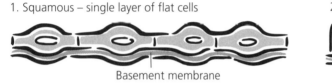

1. Squamous – single layer of flat cells

Basement membrane

2. Cuboidal – single layer of cube-shaped cells

Basement membrane

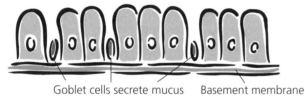

3. Columnar – single layer of tall cells together with goblet cells

Goblet cells secrete mucus   Basement membrane

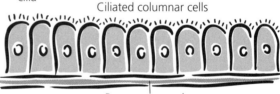

4. Ciliated – squamous, cuboidal or columnar cells with cilia

Ciliated columnar cells

Basement membrane

2. Compound epithelium is comprised of many layers of cells and comes in two different varieties:

- **Stratified** – many layers of squamous, cuboidal or columnar cells, which form a protective surface. The cells are either dry and hardened or wet and soft. Hardened cells are **keratinised** which means that the cells have dried out to form the fibrous protein **keratin**. Soft cells are non-keratinised. Examples of dry cells include the upper layers of the skin, the hair and the nails. Examples of wet cells include the lining of the mouth and the tongue.

Compound epithelium

Stratified

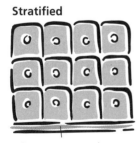

Basement membrane

Transitional

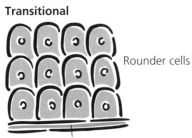

Rounder cells

Basement membrane

● **Transitional** – similar in structure to non-keratinised stratified epithelium except that the cells tend to be large and rounded rather than flat. This allows the tissue to stretch and forms structures like the bladder that need to be expandable.

Both simple and compound epithelium attach to connective tissue for support. The point of attachment between the two types of tissue is known as the **basement membrane.**

Connective tissue

Connective tissue is either solid, semi-solid or liquid. It consists of eight different types of tissue: **areolar**, **adipose**, **lymphoid**, **elastic**, **fibrous**, **cartilage**, **bone** and **blood**.

1. Areolar tissue is semi-solid in structure, permeable (allowing substances to pass through it) and is found all over the body connecting and supporting other tissue. It consists of a loose arrangement of the protein fibres **collagen**, **elastin** and **reticulin** which provide strength, resilience and support respectively.

2. Adipose tissue is also called fatty tissue and is semi-solid in structure. It is present wherever areolar is located forming an insulating layer under the skin, which helps to retain body heat.

3. Lymphoid tissue is semi-solid tissue which contains cells that help to control disease by engulfing bacteria. Lymphoid tissue forms the parts of the body systems that are involved with the control of disease.

4. Elastic tissue is semi-solid, forming elasticated fibres that are able to stretch and recoil when necessary e.g. the stomach.

5. Fibrous tissue is strong and tough and is made up of connecting fibres of the protein **collagen**. It forms connections within the body e.g. **tendons** which connect muscles to bones and **ligaments** which attach bone to bone.

6. Cartilage tissue is solid in structure and provides the body with connection and protection in the form of *hyaline cartilage* found between bones at joints, *fibrocartilage* found as discs between the bones of the spine and *elastic cartilage* found in the ear.

7. Bone tissue is solid in structure forming tough, dense **compact** bone and slightly less dense **cancellous** bone, which together form the skeletal system.

8. Blood is fluid in structure containing 55 per cent **plasma** and 45 per cent **cells**. The plasma forms the

Fascinating Fact

Liposuction is a surgical procedure that involves suctioning out fatty tissue from certain areas of the body e.g. hips and thighs.

bulk of the fluid structure of blood with the cells forming the protective and connecting functions.

Muscular tissue

Muscular tissue provides the body with movement and consists of three different types of tissue: **skeletal**, **visceral** and **cardiac**.

1. Skeletal muscular tissue is striated in appearance and provides the body with *voluntary* movement e.g. the movements involved in walking.

2. Visceral muscular tissue is smooth in appearance and provides the body with *involuntary* movement e.g. the movement of food going through the digestive system.

3. Cardiac muscular tissue provides the movement of the heart in the form of the *heartbeat*.

Nervous tissue

Nervous tissue is arranged in bundles of fibres and is made up of two kinds of cells: **neurons** and **neuroglia**. Neurons are long, delicate cells, which receive and respond to stimuli through their various parts, and neurolgia are cells that support and protect the neurons.

Remember

Some glands of the body are dual functioning and are both endocrine and exocrine e.g. pancreas, ovaries and testes.

Organs and glands

In many areas of the body, different types of tissues are joined together to form **organs** and **glands.** Organs have a specific structure and function, and consist of two or more different types of tissue e.g. heart, lungs, liver, brain and stomach. Glands are formed from epithelial tissue and produce specialised substances. There are two types of glands: **endocrine** and **exocrine**. Endocrine glands are classified as *ductless* because they pass their substances directly into the blood e.g. hormones are passed directly from the endocrine glands into the blood stream. Exocrine glands pass their substances into **ducts** (small tubes) e.g. sweat passes from a sweat gland and through a duct before reaching the surface of the skin.

Body systems

Groups of associated organs and glands that have a common function form body *systems* e.g. integumentary, skeletal, muscular, respiratory, circulatory, digestive, genito-urinary, nervous and endocrine systems.

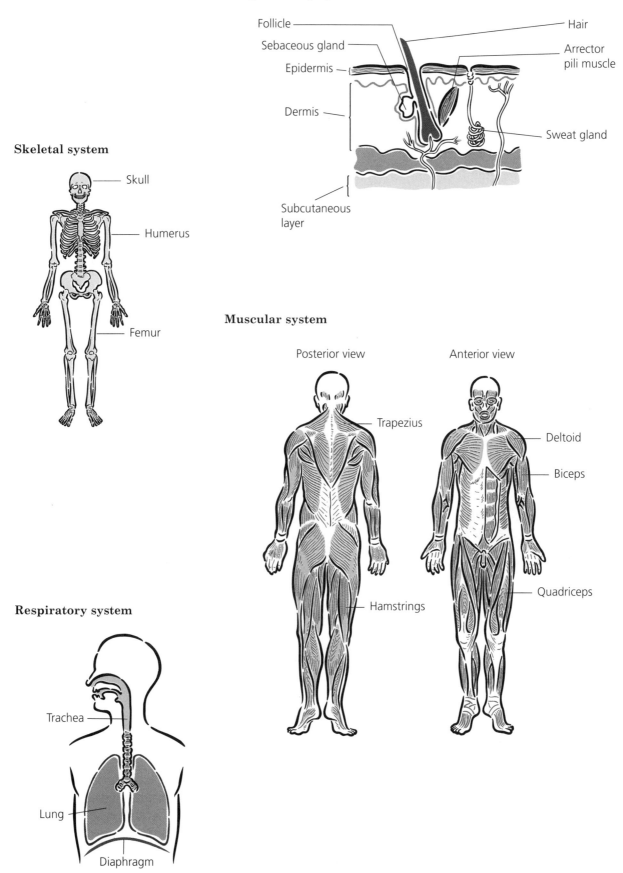

Integumentary system

Follicle

Sebaceous gland

Epidermis

Dermis

Subcutaneous layer

Hair

Arrector pili muscle

Sweat gland

Skeletal system

Skull

Humerus

Femur

Muscular system

Posterior view

Anterior view

Trapezius

Deltoid

Biceps

Hamstrings

Quadriceps

Respiratory system

Trachea

Lung

Diaphragm

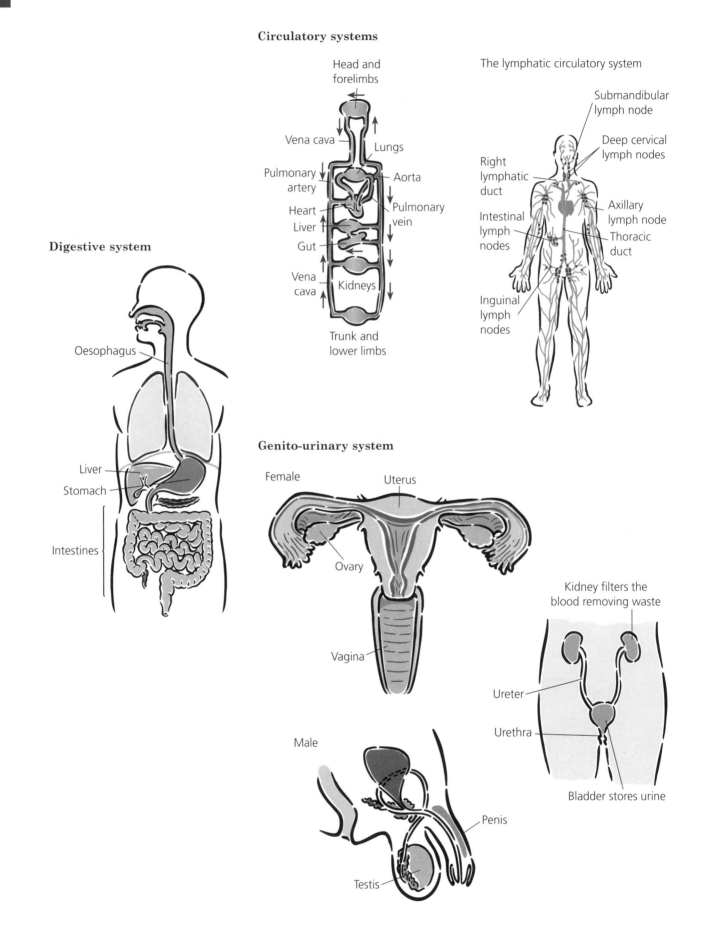

Circulatory systems

Head and forelimbs

Vena cava

Lungs

Pulmonary artery

Aorta

Heart

Pulmonary vein

Liver

Gut

Vena cava

Kidneys

Trunk and lower limbs

The lymphatic circulatory system

Submandibular lymph node

Deep cervical lymph nodes

Right lymphatic duct

Axillary lymph node

Intestinal lymph nodes

Thoracic duct

Inguinal lymph nodes

Digestive system

Oesophagus

Liver

Stomach

Intestines

Genito-urinary system

Female

Uterus

Ovary

Vagina

Kidney filters the blood removing waste

Ureter

Urethra

Bladder stores urine

Male

Penis

Testis

Endocrine system

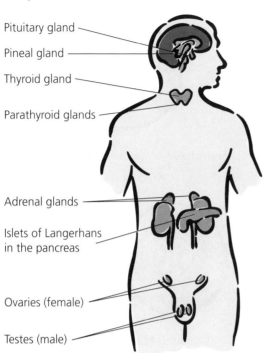

Pituitary gland
Pineal gland
Thyroid gland
Parathyroid glands

Adrenal glands

Islets of Langerhans
in the pancreas

Ovaries (female)

Testes (male)

Nervous system

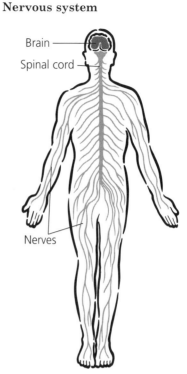

Brain
Spinal cord

Nerves

Organisms

Finally, an **organism** is formed as all the body systems function with one another producing a living human being.

Functions of cells

The function of a living human being begins with the functions of individual cells and includes: **reproduction**, **metabolism**, **respiration**, **excretion**, **movement** and **sensitivity**.

Reproduction

Reproduction of cells occurs in two ways: **meiosis** and **mitosis**. Meiosis is the process whereby a new organism is produced and mitosis is the process whereby a cell divides to form two *daughter* cells for growth and repair.

- Meiosis – a new organism is produced by the fusion of a sperm from the male with an egg from a female. There are only 23 chromosomes present in the egg and the sperm, half the number in other cells. When fertilisation takes place, the egg and the sperm fuse together to form a single complete cell known as a **zygote** with 46 chromosomes (23 chromosomes from

The stages of mitosis

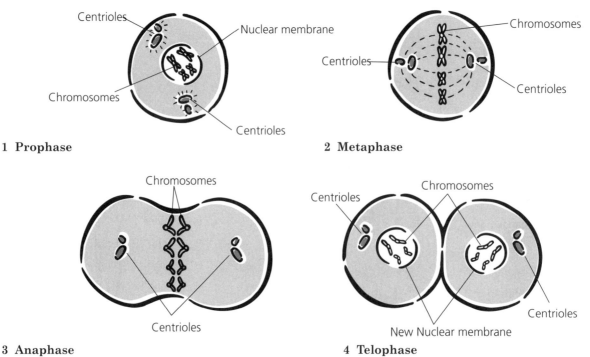

1 **Prophase**   2 **Metaphase**

3 **Anaphase**   4 **Telophase**

each parent). The zygote is then able to reproduce itself by simple cell division (mitosis) to form the embryo, the foetus and eventually the fully formed person. During this development, the cells start to specialise, with some cells becoming muscle cells, others becoming bone cells etc.

● Mitosis – the simple division of cells is a process that continues throughout life replacing old cells as they become damaged and die. The life span of most individual cells is limited and they need to be replaced if life is to continue. Mitosis is also responsible for the replication of cells needed for growth and for this reason, the process is faster in children and slows down with age.

There are four stages of mitosis: **prophase**, **metaphase**, **anaphase** and **telophase**.

1. During prophase, the single pair of centrioles present in the cell replicate themselves with each pair moving to an opposite end of the cell. At the same time the chromosomes in the nucleus begin to form into visible pairs as the membrane surrounding the nucleus starts to break down.

2. During metaphase, the chromosomes arrange themselves in the centre of the cell midway between the centrioles as the protective membrane surrounding the nucleus disappears.

Remember

Cancer can develop when changes occur within the cell causing uncontrollable mitosis.

3. During anaphase, there is further separation of the centrioles. The individual chromosomes start to move in opposite directions following the centrioles. The cytoplasm forms a 'waist' and the cell starts to constrict in the middle. This process is known as **cytokinesis.**

4. During telophase the cytoplasm continues to constrict until two identical 'daughter' cells are formed. A new protective membrane forms around the chromosomes and each cell contains a single pair of centrioles.

The organelles within these daughter cells are incomplete at cell division but they develop as the cell matures, known as **interphase**, before dividing again.

The frequency of cell division depends on the cell type e.g. skin cells reproduce more quickly than bone cells.

Metabolism

Metabolism refers to the chemical reactions that take place in a cell and may be classified as **catabolism** and **anabolism**:

- Catabolism refers to the chemical reactions that take place within the cell to break down the nutrients received into their simplest forms for the production of energy and subsequent waste products.

- Anabolism refers to the chemical reactions that take place within the cell to produce new parts of the cell structure.

Respiration

Respiration is the controlled exchange of the gases oxygen and carbon dioxide by the cell to activate the energy needed for the cell to function.

Cells are bathed in a fluid known as **interstitial** or **tissue fluid**. This fluid allows the interchange of substances between the cells and the internal transportation systems e.g. blood. The blood carries oxygen from the respiratory system (Chapter 5) and nutrients from the digestive system (Chapter 7) to the cells and these are absorbed through the cell membrane in five different ways: **diffusion, osmosis, dissolution, active transport** and **filtration**.

1. Diffusion is a process whereby small molecules such as oxygen and carbon dioxide pass easily through the tiny holes in the semi-permeable cell membrane.

2. Osmosis is a process whereby water transfers across the cell membrane when the concentration or pressure is greater on one side than the other until the concentration on both sides is equal.

3. Dissolution is a process whereby fatty substances that are too big to pass through the cell by diffusion, dissolve into the cell membrane.

4. Active transport is a process whereby energy is produced by the cell to create a carrier substance. This carrier substance actively transports other substances that are too big to pass through by diffusion and cannot dissolve into the membrane, from one side to the other.

5. Filtration is a process whereby the movement of water and soluble substances occur across the cell membrane caused by the difference in pressure. The sheer force of the weight of the fluid pushes against the cell membrane moving it through the cell.

Movement of substances through a cell

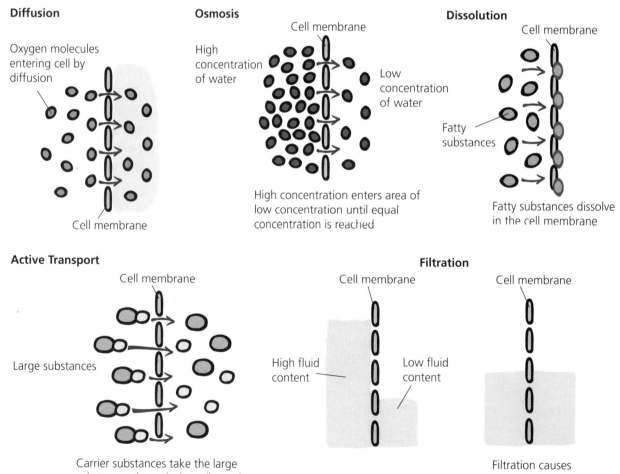

Diffusion

Oxygen molecules entering cell by diffusion

Cell membrane

Osmosis

Cell membrane

High concentration of water

Low concentration of water

High concentration enters area of low concentration until equal concentration is reached

Dissolution

Cell membrane

Fatty substances

Fatty substances dissolve in the cell membrane

Active Transport

Cell membrane

Large substances

Carrier substances take the large substances through the cell membrane

Filtration

Cell membrane

High fluid content

Low fluid content

Cell membrane

Filtration causes equal fluid content

Excretion

Excretion is the removal of the waste products from the cell. Waste products are produced as a result of respiration and metabolism and need to be removed from the cell. The removal of waste products from the cell works in the same way as absorption of nutrients into the cell.

Movement

Movement is a function of the part or the whole of certain cells e.g. the tiny hairs (cilia) of some cells move and the whole of a blood cell moves around the body.

Sensitivity

Cells are sensitive to stimuli. This means that they are able to 'pick up' messages from other parts of the cell or other parts of the body and activate a suitable response.

The functions of the cells are replicated on a larger scale in the formation of tissues, glands, organs and systems, all of which will be explored in detail as we travel through the body.

Common conditions

Disease and illness are the result of the breakdown of cells. This has a 'knock on' effect on the tissues, organs and systems of the body as the condition progresses with the potential to affect the whole organism that is the human body.

The breakdown of cells may be caused by a number of factors including: **genetic** e.g. inherited conditions, **degenerative** e.g. ageing, *environmental* e.g. extreme temperatures and *chemical* e.g. pesticides.

Other causes include: **microbes** e.g. **viruses**, **bacteria** and **fungi** as well as **parasites** such as *worms*, *insects* and *mites*.

- Viruses can only survive in living cells, which they invade and then they multiply in number, causing infection e.g. cold sores (herpes simplex).

- Bacteria are capable of surviving outside of the body and are classified as either **pathogenic** or *non-pathogenic*. Pathogenic bacteria are harmful and cause disease e.g. impetigo, and non-pathogenic are harmless helping to keep the body healthy e.g. harmless bacteria live on the surface of the skin and help to protect it.

Tip

The investigation of disease is known as **pathology.**

Fascinating Fact

It is believed by some people that the whole body cannot be completely well unless all of its parts are well and that we should treat the cause not the symptom.

Fascinating Fact

Antibiotics are used to fight bacterial infections. However, overuse can encourage the development of stronger bacteria that resist the effects of the antibiotic. When this occurs, a different antibiotic must be used.

Remember

Strict hygiene precautions are necessary when undertaking beauty and holistic treatments to avoid any possibility of cross infection.

Remember

These symptoms all form the basis of conditions that are likely to be contra indicated to treatments.

- Fungi need other cells to survive and can also be both pathogenic and non-pathogenic. Pathogenic fungi produce infectious conditions e.g. athletes foot and non-pathogenic fungi can be used to produce antibiotics e.g. penicillin.
- Worms, insects and mites are classed as **infestations** e.g. threadworms, fleas, lice, itch mites.

Microbes are **infectious** and may be passed from person to person in a process called *cross infection*. Cross infection may occur as a result of personal contact e.g. touch etc. or by contact with an infected instrument e.g. make up brush etc.

The processes of disease and illness can take on a variety of forms or symptoms including: **inflammation**, fever, **oedema**, allergic reaction and **tumours**.

- Inflammation – redness, heat, swelling, pain and loss of function.
- Fever – an increase in normal body temperature.
- Oedema – swelling as a result of an excessive amount of fluid in the tissues.
- Allergic reaction – extreme sensitivity to a substance (allergen) that is normally harmless.
- Tumours – abnormal growth of tissue, which may be **benign** (not serious) or **malignant** (tendency to become progressively worse resulting in death).

Disease and illness may be classified as either **local** or **systemic**, **congenital** or **acquired** and **acute** or **chronic**:

- Local – a condition affecting one part or a limited area of the body.
- Systemic – a condition affecting the whole of the body or several of its parts.
- Congenital – a condition present in the body from the time of birth.
- Acquired – a condition that has developed since birth.
- Acute – a condition that is sudden and severe but short in duration.
- Chronic – a condition that is long in duration.

Each of the following chapters will describe the common conditions pertaining to each body system.

System sorter

CELLS

Skeletal — Muscular — Respiratory

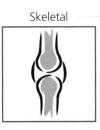

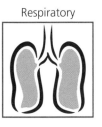

Specialised bone cells form a dense solid connective tissue that develops into the structure of the skeletal system.

Muscle cells form 3 different types of tissue. 1. Skeletal which controls voluntary movements. 2. Smooth which controls involuntary movements and 3. cardiac which controls movement of the heart.

Parts of the respiratory system like the nose contain ciliated columnar cells which form the epithelial lining and trap unwanted particles preventing them from entering the body.

Integumentary

Circulatory

The cells of the surface of the skin, the hair and the nails form stratified, keratinised epithileum. This means that they are formed in layers, are hard and dry and formed from the protein KERATIN.

Blood cells form fluid connective tissues that are able to move around the body transporting substances to and from the cells.

Mucus is produced along the linings of the digestive tract by goblet cells in the epithelial tissue. This mucus helps the flow of nutrients and waste through the system.

Endocrine glands are made from epithelial tissue and are classified as such because they secrete substances (hormones) directly into the blood.

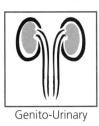

The lining of the bladder is formed from transitional epithelial tissue which is contractable allowing the bladder to expand when full of urine and deflate when empty.

Specialised cells known as neurons and neuroglia form nervous tissue which enables the body to respond to external and internal stimuli because of their function of sensitivity.

Digestive

Endocrine

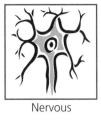

Genito-Urinary

Nervous

The human body is made up of millions and millions of cells. Specialised cells group together to form tissue and groups of tissue work together to form glands and organs. Associated organs and glands form body systems which all work together to create the organism that is the human body!

Holistic harmony

Fluid

The human body is made up of a large percentage of water – up to 75 per cent of the total body weight. Most of this water is found in cells and is known as **intracellular fluid**. The rest is found in body fluids such as blood and mucus and is known as **extracellular fluid**. The amount of water present within the body depends on the amount of adipose (fatty) tissue present, as well as sex and age. Fat cells do not contain water and as a result lean people have a greater proportion of water to total body weight compared to those people with more layers of adipose tissue. In addition, females generally carry more fat than males. The proportion of water within the body also decreases with age, with infants having the highest amount of water per body weight. Eating and drinking provides most of the water needed by the body and this is known as *performed water*. Another source of water is that which is produced through catabolism and is known as *metabolic water*. The average daily water input is about one and a half litres, which equals the average amount of water lost throughout the day. Water is lost from the body through the release of urine, faeces, sweat and breathing. If the body loses more water than it takes in during the course of the day then **dehydration** occurs. The balance of fluid within the body is regulated by thirst. When water loss is greater than water intake, the resulting dehydration causes the mouth to feel dry. The brain 'picks up' on the dry feeling and interprets this in the form of the sensation of thirst. The resulting desire to drink rebalances the fluid levels within the body.

Nutrition

The body is made up of systems formed by organs (which are made up of tissue) and which take their origin from cells. The formation of the human body from cells is reliant on nutrition. Nutrition is the process which involves the taking in, breaking down and use of nutrients. Nutrients can be classified as food, of which there are six classes: fats, carbohydrates, proteins, vitamins, minerals and water. The *nutritional status* of the body refers to the condition of the body as a result of receiving and using these nutrients, with *ideal nutrition* being the goal. Ideal nutrition refers to the best possible intake and balance of nutrients to be as healthy as possible. The results of this can be seen in mental, physical and emotional well-being. Whilst a balanced diet including the six classes of nutrients is of benefit to the body generally, certain nutrients have specific benefits for certain body systems. This will be explored in each of the following chapters with a focus on the types of foods recommended for the well-being of each body system and why.

Rest

The body relies on a certain period of time during the course of each day when it can sleep. Sleep is a period of rest for both the body and mind. Whilst asleep the body is in a state of partial unconsciousness allowing most of the bodily functions to be suspended. The body needs this period of complete rest in order to 'recharge its batteries'. The amount of sleep needed is dependent on age, activity, lifestyle and stress levels, and it varies from person to person, ranging from sixteen hours for babies to five hours for the elderly. Sleep occurs in two phases; **NREMs, n**on-**r**apid **e**ye **m**ovement **s**leep and **REMs, r**apid **e**ye **m**ovement **s**leep. NREMs is deep, dreamless and makes up about 80 per cent of sleep. REMs is associated with dreaming, usually occurs three to four times a night and can last for up to an hour.

Activity

In the same way as the body needs rest, it also needs activity in order to keep it healthy. The human body contains cells, tissues, organs and systems that are responsible for movement, some of which is under our voluntary control. If we do not make use of this facility and choose to lead a sedentary lifestyle, voluntary movement will become limited. Limited physical activity has the potential to produce limited mental activity and the phrase 'if you don't use it you will lose it' refers to both body and mind. The balance between rest and activity relevant to each of the body systems will be explored in the relevant chapters.

Air

Air is a mixture of gases that form the atmosphere in which we live. It is composed of approximately 78 per cent nitrogen, 21 per cent oxygen and 1 per cent of other gases including carbon dioxide. The air also contains varying amounts of moisture, pollution, dust etc. When we breathe in, we take in the contents of the surrounding air using up approximately 4 per cent of the oxygen before breathing out. As a result of the use of oxygen, carbon dioxide is produced in the cells and our breath out contains slightly more carbon dioxide and less oxygen than the inward breath. The nitrogen levels in the air do not change. The oxygen in air is vital for maintaining life, without it we would die in a matter of minutes. However, the additional contents in the air such as pollutants can cause the body harm. Pollutants may be found in varying amounts in the air and should be avoided in excess if at all possible e.g. passive smoking involves breathing in air that has been polluted with cigarette smoke and has the potential to cause a number of problems within the body. The art of breathing is grossly under estimated and is something that we will be exploring in order to ensure that we are able to gain maximum benefit from this natural function.

Age

Ageing is a progressive failure of the body's responses to maintaining homeostasis. Individual cells that are capable of reproduction by mitosis are thought to have a limited amount of time programmed into them in which to reproduce. The slowing down and eventual halting of processes vital to life demonstrate this theory. Another factor thought to contribute to the ageing process is the effect of **free radicals**. Free radicals are the toxic by-product of energy metabolism and include pollution, radiation and certain foods. They cause damage to individual cells because of their effect on the cell's ability to take in nutrients and release waste products. As a result of both theories, ageing produces noticeable changes in the anatomy and physiology of the human body. This process of gradual deterioration has the overall effect of increasing the body's vulnerability to illness and disease as well as producing both physical and emotional symptoms that we find it hard to cope with.

Colour

Colour is a necessary part of all life. Each living cell is dependent on light for its survival and colour is contained within light. Plants need light to produce oxygen and humans need oxygen to survive. The radiant energy of sunlight provides the body with a source of nourishment that has the power to feed the physical, emotional and spiritual elements that make up a human being. Changes in light affect changes within the body e.g. the sunrise sets

Fascinating Fact

The colours of the rainbow become visible when sunlight passes through droplets of rain. Red forms on the outer edge of the rainbow as it has the longest wavelength and the lowest frequency through to violet on the inner edge of the rainbow, which has the shortest wavelength and the quickest frequency.

off a process within the body that awakens us and in contrast the sunset and the resulting loss of light, sets off another set of processes within the body that make us feel sleepy. The colours contained in light are both visible and invisible. The visible colours make up about 40 per cent of the rays of light and can be seen because of their differing vibrations and wavelengths. The visible colours include red, orange, yellow, green, blue and violet – the colours of the rainbow. When these colours merge together they form white light.

Light enters the body through the eyes and the skin. The eyes are stimulated by the light entering them and send messages to the brain, which in turn interprets the colours. The skin 'feels' the different vibrations associated with each colour. This process is mostly unconscious but can be developed through our fingertips and hands and is often referred to as 'colour healing'.

Each colour not only has an effect on the body depending on its wavelength, frequency and vibration, but is also associated with a body system or body part and we will be exploring this further with each chapter.

Awareness

Awareness of the anatomical and physiological terms will help you to develop your understanding of the human body.

Anatomy refers to structure and there are a common set of terms that are used to describe the anatomical descriptions and positions:

Anterior – front
Posterior – rear
Inferior – below
Superior – above
External – outer
Internal – inner
Supine – laying face up
Prone – laying face down
Deep – away from the surface
Superficial – close to the surface
Longitudinal – running from top to bottom
Transverse – running across
Midline – line through the centre of the body
Medial – towards the midline
Lateral – away from the midline
Distal – furthest point away from an attachment
Proximal – closest point to an attachment

Physiology refers to functions and the following terms are used to describe the study of different systems:

Histology – the cells and tissues
Dermatology – the integumentary system
Osteology – the skeletal system
Myology – the muscular system
Cardiology – the heart
Haematology – the blood
Gastro-enterology – the digestive system
Gynaecology – the female reproductive system
Nephrology – the urinary system
Neurology – the nervous system
Endocrinology – the endocrine system

Special care

Homeostasis is the condition in which the cells, tissues, glands, organs and systems of the body are working in harmony within themselves and with one another.

These combined activities provide the best possible environment for the well-being of individual cells and the maintenance of this environment is vital for the well-being of the whole body.

One of the main factors which affects homeostasis is **stress.** Stress can be **external** e.g. changes in temperature, loud noises, lack of oxygen etc. or **internal** e.g. pain, worry, fear etc. The body has many ways in which to cope naturally with the everyday routine stresses that it encounters and has clever mechanisms in place to counteract the effects. However, this constantly changing situation must be kept within certain limits otherwise an imbalance occurs. Serious imbalance caused by excessive and prolonged stress results in the body failing to cope and ill health may result.

Beauty and holistic treatments help a client to recognise the effects of stress, hopefully before it becomes too pronounced, and ongoing treatment combined with after care advice can form the basis of the special care needed to ensure that any imbalance is avoided and homeostasis is maintained.

Tip

Homeo = same and
Stasis = state

Angel advice

As therapists, it is also important that we take note of the special care needed by the body and that we 'practice what we preach'!

Treatment tracker

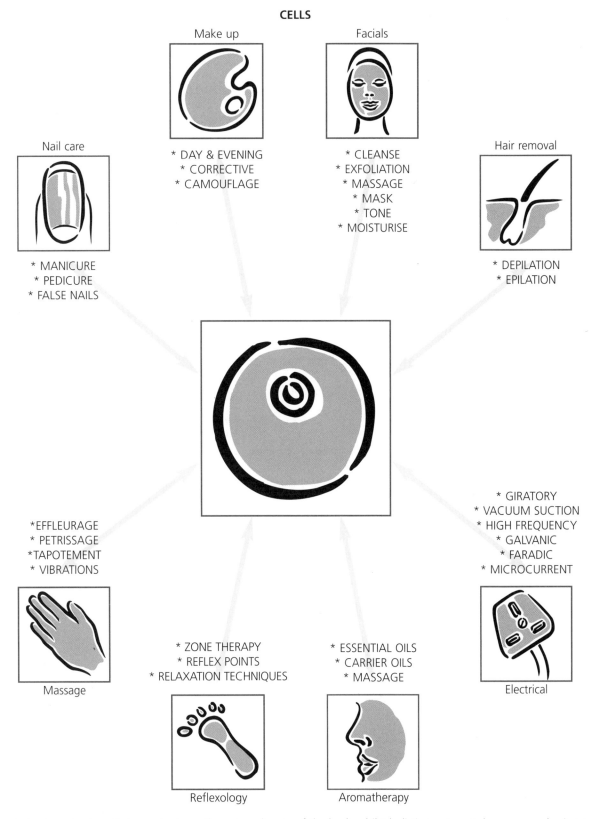

CELLS

Make up
* DAY & EVENING
* CORRECTIVE
* CAMOUFLAGE

Facials
* CLEANSE
* EXFOLIATION
* MASSAGE
* MASK
* TONE
* MOISTURISE

Nail care
* MANICURE
* PEDICURE
* FALSE NAILS

Hair removal
* DEPILATION
* EPILATION

*EFFLEURAGE
* PETRISSAGE
*TAPOTEMENT
* VIBRATIONS

Massage

* ZONE THERAPY
* REFLEX POINTS
* RELAXATION TECHNIQUES

Reflexology

* ESSENTIAL OILS
* CARRIER OILS
* MASSAGE

Aromatherapy

* GIRATORY
* VACUUM SUCTION
* HIGH FREQUENCY
* GALVANIC
* FARADIC
* MICROCURRENT

Electrical

Beauty treatments place their emphasis on the external parts of the body whilst holistic treatments have an emphasis on the internal parts of the body. It is important to appreciate that you cannot have one without the other. All treatments have the potential to have a beneficial affect on all parts of the body – from the tiniest cell to the largest body system.

Knowledge review – Cells

1 What is the meaning of anatomy and physiology?

2 What is the name given to the part of the cell described as the 'information centre' and what does it contain?

3 How many chromosomes does each cell contain?

4 What are organelles?

5 Name four organelles.

6 What is the name given to a group of cells?

7 Name the four different types of tissue.

8 What is the name given to the cells which secrete mucus?

9 Which type of tissue contains fat cells?

10 What are collagen, elastin and reticulin fibres made from?

11 What are the three different types of muscular tissue called?

12 What type of tissue is blood?

13 What do two or more tissue types produce?

14 What is the difference between meiosis and mitosis?

15 What is the difference between catabolism and anabolism?

16 Name two methods to describe the movement of substances through the cell membrane.

17 Name three microbes responsible for causing illness and disease.

18 What does contra indication mean?

19 What are the anatomical terms used to describe front, rear, away from the midline and below?

20 What is the meaning of homeostasis?

The integumentary system

2

Learning objectives

After reading this chapter you should be able to:

- **Recognise the structure of the skin, hair and nails**

- **Identify the different layers of the skin**

- **Understand the functions of the skin, hair and nails**

- **Be aware of the factors that affect the well-being of the integumentary system**

- **Appreciate the ways in which the integumentary system works with the other systems of the body to maintain homeostasis.**

Our exploration of the human body continues on the outside with the largest organ of the body, **THE INTEGUMENTARY SYSTEM**, consisting of the skin, hair and nails. This system provides the body with a waterproof protective outer covering that is resilient, flexible and contributes to our unique personal appearance. As our journey takes us through the layers of skin and its parts, we will begin to appreciate the complexity of its structure and marvel at the way in which its functions contribute to the harmonious working of the whole body.

Science scene

Structure of the skin, hair and nails

The outer most layer of the skin, the one we can see and touch, is called the **epidermis**, beneath which lies the **dermis** and finally the **hypodermis**. The hair and nails are extensions of the skin.

The skin

The skin is made up of two main layers, the dermis and the epidermis, under which lies a layer of fatty tissue called the hypodermis. The muscles, the bones and/or the organs of the body lie directly below these layers.

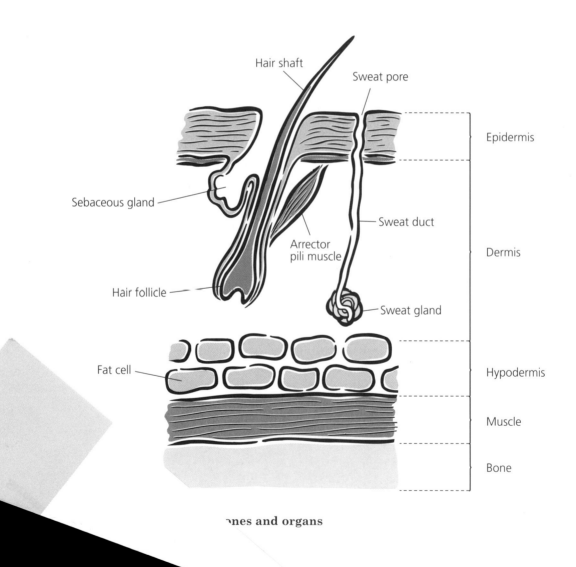

ones and organs

Tip

It is the condition of the hypodermis that contributes to the 'orange peel' appearance of **cellulite.** The distribution of fat cells is affected by the circulatory and endocrine systems and if an imbalance is experienced the skin tends to take on this 'orange peel' effect, which is difficult to treat. This is more common in women than in men and tends to be on the upper thigh area where the natural distribution of fat cells is greater, the circulation poorer and the hormone activity varied as women go through their menstrual cycle.

The hypodermis

We are going to start by looking at the deepest layer lying directly above the muscles of the body, the **hypodermis**, also known as the **subcutaneous layer** or **subcutis.** It is made up of two main types of connective tissue, **areolar** and **adipose** tissue.

- Areolar tissue forms a loose network of cells providing strength, elasticity and support. Strength to protect the underlying structures, elasticity to cope with an increase or decrease in size and support for the blood vessels and nerve endings which service this layer.

- Adipose tissue forms a network of fat cells providing the body with insulation thus keeping our inner body warm as well as acting as a source of energy. When we lose weight by reducing our calorie intake, the body uses up the fat cells stored in this layer as a replacement for the lack of food. As the fat cells decrease, we lose inches. Too much body fat caused by over eating puts pressure on the body to work harder but too little body fat also puts pressure on the body in that it loses its energy reserves and heat insulation. Therefore it is important to maintain a balance in what we eat for the energy we need in order to stay healthy.

The dermis

The layer of skin lying directly above the hypodermis is the dermis or *true skin*. It is called the true skin because it contains the structures associated with the skin and performs many of its main functions.

The cells of the dermis that connect with the hypodermis underneath form connective tissue making up two distinct layers – the deeper **reticular layer** and the more superficial **papillary layer**.

The reticular layer

Groups of cells in this layer form mainly areolar connective tissue containing various types of protein fibres including **collagen**, **elastin** and **reticulin**.

- Collagen fibres give the skin strength, resilience and a youthful appearance. These cells diminish with age, changing the skin's appearance from peach like to prune like!

- Elastin fibres give the skin elasticity enabling it to stretch in order to accommodate an increase in fat

Fascinating Fact

One square centimeter of skin holds approx. 15 sebaceous glands, 100 sweat glands, 3 m. of nerves and 1 m. of blood vessels!

cell distribution in the form of weight gain and of course pregnancy. Stretch marks occur when the elastin fibres fail to respond efficiently to this increase resulting in over stretched skin. However, we do not have to fear that skin will break as a result of too much stretching. It has the ability to withstand extreme weight gain but would prefer it to be moderate if it is to maintain this function efficiently without putting undue strain on the skin as a whole!

- Reticulin fibres provide support for the many structures held within this layer that include hair follicles, sebaceous glands, sweat glands, arrector pili muscles, nerve supply and circulatory vessels.

- Hair follicles are tube like structures within which a hair develops and grows. Hairs extend up and out onto the skin through a **pore** (a minute opening in the skin's surface). Each hair follicle has a rich blood supply feeding the cells that form the structure of the hair, as well as an active nerve supply linking it to the nervous system allowing us to detect pain when a hair is pulled. As a hair extends out of the skin, its cells disconnect from the blood and nerve supply lower down and start to die off, a process

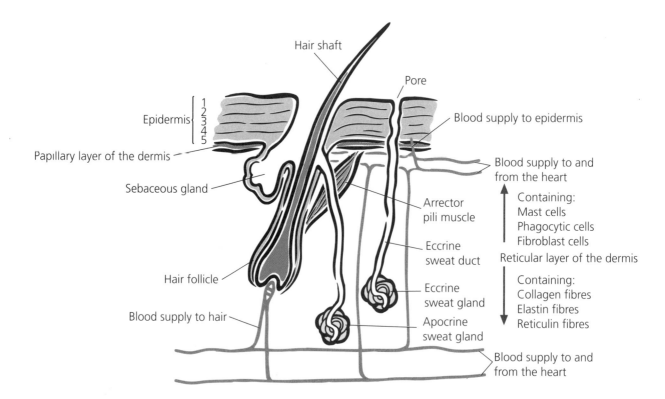

The dermis and its structures

Tip

The process of keratinisation also occurs in the top layer of skin forming the hard dead skin cells that make up the surface of the skin, as well as in the nails, forming the hard structures protecting the tips of the fingers and toes.

known as keratinisation. The cells form into stratified epithelial tissue, as they become hard, flat and dead and it is this formation of cells that allows us to cut hair without pain or bleeding!

- Sebaceous glands are generally attached to hair follicles and produce the skin's natural lubricant – an oily substance called **sebum**. Sebaceous glands are exocrine glands, formed from epithelial tissue, which secrete their substance into a duct. The hair follicles act as the ducts for the sebum produced by the sebaceous glands to travel up and out eventually reaching the skin's surface via a pore. Sebum provides lubrication for the skin and hair maintaining the skin's suppleness and the hair's lustre. As we comb or brush our hair the sebum is taken down its length. For this reason, it was once thought that brushing the hair one hundred times a night would keep hair in good condition. The longer the hair, the further the sebum has to travel down its length, and the less efficient the sebum is in lubricating the ends. Added to which, the use of heat in the form of hair driers and tongs etc. evaporates the sebum reducing its lubricating qualities and the slowing down process associated with ageing contributes to the drying out of hair making a hundred brush strokes a night inefficient and time consuming! Nowadays we replace lost or lacking sebum with conditioner! Likewise, sebum needed to lubricate the skin may be lost as a result of extreme temperatures as well as the general effects of the ageing process slowing down the activity within the glands. In order to combat this, we apply a moisturiser. Sebum is also slightly antiseptic and helps to create a vital barrier on the skin's surface helping to keep moisture in and invaders out! Sebaceous glands may become over active as a result of hormone imbalance resulting in excessively oily skin and hair. This can be most distressing for the group of people affected by the over activity in these glands. Younger clients often seek help from a therapist which may be given in the form of specific skin care i.e. the use of clay masks to absorb the excessive oil or reflexology to help to balance the body systems responsible for the over activity.

- Sweat glands are also exocrine glands and are collectively known as **sudoriferous glands** of which there are two types **eccrine** and **apocrine**.

    1. **Eccrine glands** are found all over the body but are more numerous on the palms of the hands, the soles of the feet and under the

Remember

Remember that the skin plays host to many bacteria that live on its surface.

Tip

As therapists we need to be aware that if our client has 'goose bumps' they are either cold or scared!

arms. Their function is to produce sweat, which helps to regulate body temperature by cooling the body when it gets too hot. The sweat from these glands travels to the surface of the skin via a sweat duct, which opens onto a sweat pore.

2. **Apocrine glands** are found mainly in the underarm and genital areas of the body, are inactive in children, and develop in puberty. These glands open onto hair follicles. The sweat produced from these glands is broken down by bacteria present on the surface of the skin resulting in body odour.

- Arrector pili muscles are sometimes referred to as *erector pili* muscles. They are small structures attached to hair follicles. They are comprised of smooth muscular tissue and are under involuntary control responding automatically to changes in temperature. The tiny muscles contract as the body temperature drops from 36.8°C, pulling the skin into 'goose bumps' while lifting the corresponding hairs, trapping the warm air beneath them in an attempt to keep the body warm. These muscles will also contract in response to extreme emotions. For example we develop 'goose bumps' when we hear a piece of music that has great meaning for us or when we see something that fills us with fear!

- Nerve supply involves a complex network of nervous tissue that links the skin with the nervous system. The skin is a *sensory organ* which means that it is able to 'pick up' on any external stimuli e.g. pain, pressure or changes in temperature, and respond accordingly.

- Circulatory vessels are situated throughout this layer linking the skin with the blood and lymph supply of the body. Blood vessels are responsible for 'feeding' the cells with oxygen brought in from the respiratory system (Chapter 5) and nutrients and water from the digestive system (Chapter 7). This 'food' is vital for cellular respiration (absorption of oxygen) and cellular metabolism (the production of energy) for the growth, development, repair and renewal of the skin. Lymph vessels help to release the waste that collects as a result of this activity.

In addition to these structures, the reticular section of the dermis also contains:

- **Fibroblast cells**, which are responsible for forming new fibrous tissue e.g. collagen.

- **Phagocytic cells**, which help to defend the body by destroying bacterial invaders, and
- **Mast cells**, which produce **histamine** when the skin is damaged or irritated, allowing more blood to flow to the area for assistance.

Above the reticular section of the dermis lies the thinner papillary section, which connects with the epidermis and the hair follicle. It is formed from connective areolar tissue providing these structures with support together with a blood and nerve supply for 'feeding' (cellular respiration, metabolism) and 'feeling'!

The epidermis

The uppermost section of the skin, the epidermis, is made up of five main layers of cells which collectively form stratified epithelial tissue.

The individual layers of cells are referred to as **stratum**.

We start at the deepest layer, the **stratum germinativum** or **basal layer**, in which the cells are very active receiving a rich blood supply from the underlying papillary layer of the dermis. The cells of this layer are cuboidal and are able to reproduce themselves by mitosis (simple cell division). As these cells produce new cells, the old cells are pushed up towards the skin's surface, changing in composition as they enter each of the next four layers, until they are finally shed or **desquamated**. This is a constant process, which takes approximately one month to be completed.

Remember

Remember that stratified refers to layers, epithelial refers to cells that form a protective lining or covering and tissue refers to groups of cells.

Remember

Desquamation means the natural exfoliation of dead skin cells.

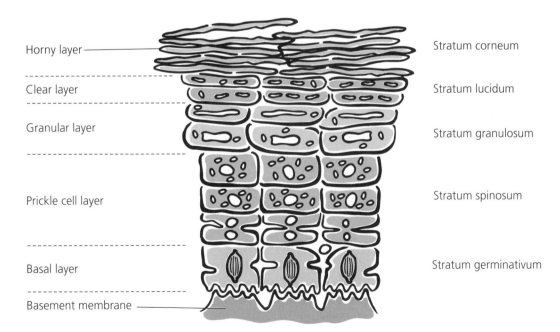

Horny layer — Stratum corneum

Clear layer — Stratum lucidum

Granular layer — Stratum granulosum

Prickle cell layer — Stratum spinosum

Basal layer — Stratum germinativum

Basement membrane —

Layers of the epidermis

There are three main types of cells present in this layer:

- Most of these cells are called **keratinocytes** and produce a protein called **keratin**.

- To assist with the protection of the body from external invasion, this layer of skin contains *Langerhan cells*, which absorb foreign bodies passing them deeper into the skin for removal via the circulatory systems.

- Star shaped cells called **melanocytes** produce the colour pigment **melanin**. Melanin is produced in abundance in dark skins providing added protection and is totally lacking in albino skins, which are pale and vulnerable as a result. The ultraviolet rays of the sun activate increased production within the melanocytes causing the additional melanin to rise, darkening the skin and producing a suntan. The melanocytes cannot cope with excessive exposure to the sun's rays, at which point the skin will burn. The skin's link with the nervous system will alert the body to this by providing the feeling of pain. Mast cells in the dermis will produce histamine in response to skin irritation, which will in turn bring more blood to the area, and the skin will go pink. It is up to us to respond to these messages or suffer the consequences, which can in the long term prove fatal with the development of skin cancer.

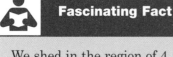

Fascinating Fact

We shed in the region of 4 per cent of our total skin cells every day which is approx. 18kg. of skin in a lifetime, which in turn makes up a large percentage of household dust!

Remember

Skin colour is also determined by the blood and its contents as well as levels of carotene, a yellow pigment stored in the skin from the food we eat e.g. carrots, tomatoes and fats.

Nails develop from this layer of the epidermis as clear, hard layers of keratinised cells providing additional protection to the extremities of the body. The keratin in nails is hard keratin compared with the soft keratin of the skin. Cells containing hard keratin are more durable and do not desquamate.

As mitosis takes place in the stratum germinativum, the old cells are pushed upwards forming the next layer, *stratum spinosum* or **prickle cell layer**, consisting of eight to ten layers of cells. Mitosis still takes place in the lower layers, but in the upper layers, keratinocytes begin the process of keratinisation changing the shape of the cells from cube-like to spiky, hence the term prickle cell; it is through these spikes that the melanin is able to enter this layer. These cells receive some nourishment from the blood flowing in the spaces between the cells.

The cells continue to move upwards forming the **stratum granulosum** or **granular layer** made up of two to five layers of diamond shaped cells, which become flat, hard and granule like as the completion of keratinisation takes place. The destruction of melanin by **enzymes** starts to take place

Remember

An enzyme is a protein produced in a cell that is capable of speeding up a chemical reaction for which it is responsible.

in this layer of the skin making the cells transparent as they reach the next layer.

As a result, this layer is called *stratum lucidum* or **clear layer** and tends to be thickest on the soles of the feet and palms of the hand as the cells contribute to the waterproofing function of the skin. The skin is able to absorb some water but the skin controls the amount, otherwise we would get out of a bath and find the bath empty!

The final layer, the *stratum corneum* or **horny layer**, consists of up to 25 layers of flat, hard cells that provide the visual appearance of the skin, although the cells of the underlying layers contribute to their overall condition. A good blood supply to the deepest layer will ensure efficient cell reproduction, which will be seen in good surface cells when they eventually reach the top a few weeks later. Because of the constant renewal of cells at the deeper levels, these surface cells are being shed continuously. If this process is hindered, the skin cells will develop a thick, hard layer that is unattractive and uncomfortable.

Think about the hard skin, which develops on pressure points of the feet for example. A beauty therapist will remove this during a pedicure with the aim of producing smooth, soft feet as a result. A holistic therapist, on the other hand, may view the situation differently and link the areas of hard skin with the areas of the body represented in the feet. In this case the hard skin is believed to be the body's way of protecting those particular organs that may be stressed. For example, the skin over the middle of the ball of the foot represents the lungs. Clients suffering with asthma may find that they have a build-up of hard skin in this area.

At this point it is useful to look in more detail at the structures which protrude from the skin. Although their functions are essentially of an external nature, their structure begins within the skin itself so an understanding of the way in which they link with one another is an important factor in the study of this system as a whole.

Fascinating Fact

An average adult usually has about 100 000 head hairs!

Hair

Hair develops from the hair follicle and adds to the protective functions of the body.

Hair is present on the body prior to birth developing as **lanugo hair**. This hair is lost just before or shortly after birth and is replaced with **vellus hair**. Areas of the body requiring additional protection develop **terminal hair**.

- Lanugo hair is soft, fine hair covering the body.
- Vellus hair is soft and downy covering the whole body except for the palms of the hands, soles of the feet, the lips and parts of the genital areas.
- Terminal hair is the coarse hair of the scalp, inside the ears, the eyebrows and eyelashes, underarm and pubic regions protecting the vulnerable areas which lie below them i.e. the brain, the eyes, the glands under the arms and the genitals. It is also present on the faces and chests of men.

Terminal hair growth and patterns of growth vary with race:

- Caucasian European hair is generally straight or loosely waved with medium amounts of facial/body hair.
- Latin hair type is coarse, straight and dark with heavy facial/body hair growth.
- Eastern hair type is coarse and straight with light facial/body hair growth.
- Afro-Caribbean hair type is tightly curled and woolly with light facial/body hair growth.

Hair growth is reliant on hormonal balance and changes occur when this balance is altered. Hormonal changes are responsible for the development of terminal hair from vellus hair, initiating stronger, coarser hair growth in certain areas of the body at certain times of our lives (Chapter 10).

- **Puberty** changes hair growth in teenagers.
- **Pregnancy** can activate hair growth making it thicker in certain areas of the body i.e. abdomen.
- **Menopause** can activate male characteristic hair growth in women.

Other factors which affect hair growth may include excessive stress levels, heredity and illness, with terminal hairs growing in undesirable areas as a result.

Social factors determine an individual's need to remove this hair, with different cultures having differing views on the desirability of hair growth!

The growth of lanugo and vellus hair are from the sebaceous glands situated in the dermis; terminal hair growth begins at the base of the hair follicle, the hair bulb where dermal papillae from the papillary layer of the dermis offer the hair-producing cells a rich supply of blood and nerves.

The hair in its follicle

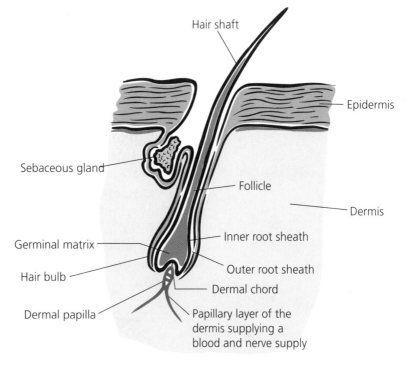

- The blood supplies the cells with nutrients from the digestive system and oxygen from the respiratory system to 'feed' the cells for growth and development as well as hormones from the endocrine system to determine the actual pattern of hair growth.
- The nerves enable the body to detect pain when hair is pulled!

The lower portion of the **hair bulb** contains the matrix or hair root. The cells within the matrix reproduce new cells by mitosis, pushing old cells upwards in much the same way as the cells in the layers of the epidermis.

Surrounding these cells is the *root sheath*. This is made up of an inner portion and an outer portion.

- *Inner root sheath* – cells inter link with the hair cells helping to secure the hair within the follicle.
- *Outer root sheath* – this is formed from the stratum germinativum and allows the renewal and development of hair cells to take place.

As the hair cells are pushed up through the bulb, they become keratinised and develop into hard layers of dead cells resulting in a strand of hair, or *hair shaft*, which can be seen on the surface of the skin as it emerges from the follicle out of a follicular pore.

Most follicles produce one hair each, which is individually made up of three layers, the **medulla**, the **cortex** and the **cuticle**:

- **Medulla** – innermost layer made up of loose cells. Not always present in all hairs or even within the length of the same hair thus giving a difference in texture and sheen.
- **Cortex** – keratinocytes and melanocytes are found in this layer determining the strength and colour of hair.
- **Cuticle** – outermost layer made up of flat overlapping cells. Within the hair follicle these cells inter link with the cells of the inner root sheath, and outside of the body, these cells provide a protective coating to the hair.

Layers of the hair

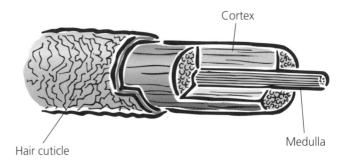

Cortex

Hair cuticle

Medulla

Individual hairs can grow up to half an inch a month but they do not grow continuously. They have a growth cycle, which takes them through three stages:

Fascinating Fact

Hairs on the head may last up to six years before falling out yet individual eyelashes fall out after only six months!

1. **Anagen** – this is the first stage of hair growth and is initiated by the endocrine system, releasing a hormone into the blood stream and sending it to the outer root sheath. The outer root sheath is then stimulated into activity and the dermal cord is produced linking the hair follicle at the dermal papilla. A new hair develops from cells in the matrix section of the hair bulb at the base of the follicle, and the dermal cord provides the means for the hair cells to receive nourishment from the dermal papilla to allow mitosis to take place.

2. **Catagen** – the second stage of growth involves the hair cells passing up through the follicle. The hair is now fully grown and as it does not need the matrix anymore, it detaches from it. The remaining parts break down and become dormant until a new hair is produced.

3. **Telogen** – the third stage of growth is the final phase of the cycle. The hair dries up and eventually falls out leaving room for the whole cycle to begin again!

The stages of hair growth follow a continuous pattern involving the following:

1. Action – **a**nagen stage, which can last for up to six years.

2. Change – **c**atagen stage, which takes place over a few days.

3. Termination – **t**elogen stage, which lasts for varying lengths of time and is dependent on hormone release to activate the new growth associated with anagen. Sometimes a new hair develops before the old hair has been released.

Stages of hair growth

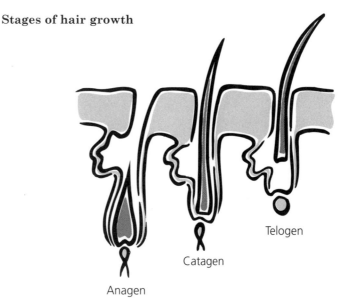

Telogen

Catagen

Anagen

Fascinating Fact

We shed in the region of 100 hairs from our heads each day as new hairs are constantly being developed

Individual hair follicles pass through these stages at different times, therefore some old hairs are always falling out. The majority of hairs will be at the anagen stage with the least amount of hairs being at the telogen stage at any one time.

Hair may also be removed using semi-permanent methods i.e. **depilation** or permanent methods i.e. **epilation**.

When a hair is removed during the anagen stage of growth, it will come away with the bulb and inner root sheath intact. Hair will also take longer to grow back because the natural cycle has been disrupted.

When a hair is removed during the catagen stage, it will appear with a black bulb at its base and a new hair will grow back fairly quickly.

During the telogen stage, the hair is ready to fall out naturally and will have lost the black bulb and natural regrowth of hair will follow quickly.

Nails

The other external structures of the skin are the nails that develop from cells within the epidermis, and protrude to cover the ends of the fingers and toes as hardened 'plates', which offer added protection to the extremities of the body.

Nails do not have a growth cycle like hair; instead, they grow continuously throughout life starting from the third month of foetal development.

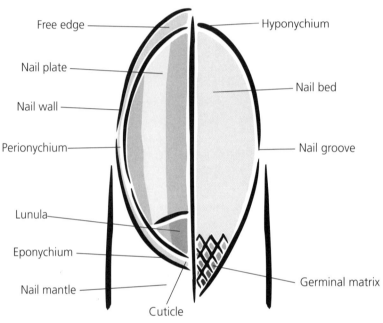

Free edge — Hyponychium
Nail plate — Nail bed
Nail wall — Nail groove
Perionychium — Lunula — Eponychium — Nail mantle — Cuticle — Germinal matrix

Structure of the nail (vertical view)

The parts of the nail include:

Germinal matrix

This is the root of the nail, situated in the stratum germinativum; it forms the cells which will eventually produce a nail, which subdivide through mitosis pushing old cells upwards towards the **lunula**.

Lunula

The cells become visible collectively as a half moon shape at the base of the nail and gradually harden through keratinisation as they push further upwards forming the nail plate.

Nail plate

This consists of three layers of clear cells which have developed with the layers of the epidermis and form an extension of the skin at the ends of the fingers and toes.

The cells are now dead, filled with hard keratin, they do not desquamate, and they are held together in longitudinal layers with moisture and fats.

Nail bed

This lies directly below the nail plate and is responsible for securing the nail to the finger or toe. The cells of the surface of the nail bed interlock with the cells of the underside of the nail plate forming a secure bond. The end of the nail plate, which extends over the nail bed, is known as the **free edge**.

Nail cuticle

This develops as an extension of the stratum corneum creating the **nail fold**. It starts as the **eponychium** at the lunula, attaching itself to the nail plate and forming a joint, which protects the underlying germinal matrix against damage and invasion. It surrounds the nail plate as the **peronychium**, extending under the top of the nail plate as the **hyponychium**, it offers protection to the nail bed and stops particles from getting under the nail plate.

Nail grooves

The nail is guided up the fingers and toes by grooves at the sides, which help to keep it on 'track'.

Nail wall

This is the skin which covers the sides of the nail plate protecting the nail grooves.

Nail mantle

This refers to the skin lying directly above the germinal matrix.

The new cells formed at the germinal matrix and nail bed are serviced with a rich blood and nerve supply from the papillary layer of the dermis.

- A good blood supply ensures a good supply of nutrients and oxygen vital for the successful development of the cells, which will eventually form the nail plate. If the blood circulation to the nail bed is good, the nails will appear pink.
- The nerve supply ensures that pain is experienced if the nails are pulled out from the matrix or up from the nail bed – an effective method of torture! OUCH!

Remember

The areas of the lunula and the mantle feel tender to touch. This is due to the formation of new cells below and highlights the need for care when treating this area to prevent damage. If the germinal matrix suffers extreme damage, this can affect the development of new cells resulting in permanent loss or deformity of the nail plate. Care should be taken therefore when performing cuticle work so as not to aggravate the underlying structures.

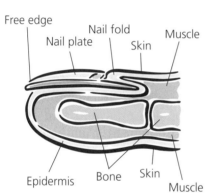

The structure of the nail (horizontal view)

Remember

Remember that the nail plate itself is dead so will not bleed or hurt when cut, just like the hair, and as with all dead cells, they receive neither a blood nor a nerve supply. They are designed purely for protective purposes although they also serve as decoration!

Nails continue to grow effectively in a healthy body and are affected by illness, stress and heredity in much the same way as hair and skin.

Functions of the integumentary system

Having gained an awareness of the structures that make up this fascinating system, we can now examine its functions.

The integumentary system has seven main functions which all contribute to the general health of the whole body in much the same way as the skin relies on other body systems to maintain its own level of well-being. These functions include protection, temperature regulation, **absorption**, **secretion**, **excretion**, sensation and production of vitamin D.

Protection

The skin protects the body like a living suit of armour, keeping vital organs safely *in* and environmental enemies *out*! The skin protects the body in the following ways:

Remember

pH is a measurement of acidity or alkalinity and ranges from 0–14. Seven is neutral with anything above being alkaline and below acidic.

Remember

Excessive exposure to the sun will result in premature ageing of the skin.

- **Acid mantle** – a protective covering to the skin formed by the combination of sebum, sweat and dead skin cells; it helps to maintain a slightly acidic pH of between 4.6 to 6 depending on the area of the body. Using products that are too harsh will break down the acid mantle leaving the skin vulnerable. Products suitable for the skin type will therefore help to restore it whilst a good skin care routine and a healthy body will maintain it.

- **Melanin production** – melanin is produced by melanocytes in the stratum germinativum as a direct result of exposure to the sun's rays and provides protection against damage to the underlying structures. Sunscreen will help to protect the skin further and allow longer, safer exposure to the sun's rays.

- **Fat cells** – in the form of adipose tissue deposited in the hypodermis, these form a protective 'cushion' against damage to the underlying systems, i.e. muscles, bones and organs.

- **Touch** – the skin's links with the nervous system alerts the body systems to respond i.e. we touch a hot plate and immediately pull away. The skin has produced an early warning system against further damage. The sense of touch stimulates, warns and activates!

- **Healing** – this occurs when the skin becomes damaged and broken. The blood rushes to the area where repair cells within the blood create a fine covering over the wound. The blood cells accumulate behind this covering to form a clot. The clot dries and forms a scab thus sealing the skin. Meanwhile the scab prevents blood from leaving the body and germs from getting in. New skin develops beneath the scab, which drops off once the skin is completely renewed. The body heals from the outside in – and from the inside out!

- **Hair** – develops from the hair follicles in the dermis and protrudes out onto the surface of the skin; terminal hair offers the skin additional protection in more delicate areas with the thicker growth and vellus hair growth in other areas.

- **Nails** – develop from the cells of the stratum germinativum in the epidermis and protrude out of the skin to protect the ends of the fingers and toes.

Temperature regulation

The skin contributes to maintaining normal body temperature of 36.8°C in the following ways:

- **Sweating** – as the body temperature rises, the sudoriferous glands produce sweat. The heat of the skin evaporates the sweat producing a cooling effect.

- **Vasodilation** – occurs when there is a rise in body temperature. Blood vessels dilate (widen), allowing the blood to flow near the skin's surface making the skin turn pink and losing the excess heat in the process.

- **Vasoconstriction** – occurs as body temperature falls by constricting (tightening) blood vessels, forcing the blood away from the skin's surface making the skin look pale. The blood will flow closer to the internal organs keeping them warm and raising body temperature.

- **Fat cells** – in the hypodermis help to insulate the body against excessive heat loss.

- **Goose bumps** – created by the contraction of the arrector pili muscles in the dermis also lift the hairs trapping a layer of warm air beneath them.

- We shiver as body temperature drops, stimulating activity and creating heat.

Fascinating Fact

The tips of the fingers and toes absorb more water than anywhere else in the body because of the lack of hair and sebaceous glands. This is the reason why these areas go wrinkly in the bath!

Absorption

Absorption of water and moisture takes place on the surface of the skin and is controlled by the acid mantle and the stratum lucidum, which allow enough absorption to take place to keep the cells of the epidermis hydrated. Certain substances with a particular molecular structure penetrate the skin further through diffusion (Chapter 1).

The molecules pass through the epidermis and are carried away by blood vessels circulating in the dermis. Aromatherapy oils are absorbed in this way and are capable of having an effect on the whole body as a result. HRT and nicotine patches penetrate the skin in the same way.

Hair and nails are also able to absorb water and moisture helping to keep them hydrated and in good condition. By absorbing colouring and perming products, the hair is able to change both its colour and its shape.

Secretion

Secretion is a cellular process of releasing a substance. Sebum, the skin's natural lubricant, is secreted by the sebaceous glands into a hair follicle and onto the surface of the skin. This secretion of sebum ensures healthy skin and hair which is soft and moist. There are an abundance of sebaceous glands present along the centre panel of the face and upper back and chest as well as generally all over the body except on the palms of the hands and soles of the feet. Sebum secretion is dependent on hormonal balance and can become over or under active as a result of hormone imbalance. Examples of times when the hormone levels can differ include: puberty, pregnancy and menopause, while stress, illness, age, gender and some drugs can also affect hormone levels.

Excretion

Excretion is a way of eliminating waste out of the body. The skin contributes to this process by excreting sweat. Sweat contains water together with small amounts of urea, uric acid, ammonia and lactic acid and is excreted from the eccrine and apocrine glands.

Sensation

The sensation of touch, pressure, pain and temperature are transmitted to the brain via an elaborate network of **sensory nerves** and receptors in the skin in the form of electrical impulses (Chapter 9). The brain interprets these impulses and transmits messages back to the skin via **motor nerves** in order to initiate a response. This process takes less than one tenth of a second to activate and is the key to our survival providing us with an intricate 'feel' for life. Touch triggers our memory and arouses our senses – it is the messenger of all emotions. Touch nurtures, soothes and heals as well as stimulates, invigorates and activates! The art of touch is the science of trust.

Production of vitamin D

Vitamin D is produced by the action of sunlight on the skin, which is in turn absorbed by the blood stream and used for the maintenance of bones. Vitamin D helps the body store calcium, which is needed for healthy bones (Chapter 3). The skin contains a fatty substance called **ergosterol**, which is converted into vitamin D in response to stimulation from the ultraviolet rays of the sun.

Having a knowledge of how the skin and its parts function helps us to be more respectful of the body as a whole and in turn more aware of the times when it does not function quite as we would want it to!

Common conditions

A–Z of common conditions affecting the skin, hair and nails

- ABSCESS – a pocket of local infection which is painful, swollen and contains a pus formation.
- ACNE ROSACEA – redness of the nose and cheeks characterised by the shape of a butterfly. Caused by dilation of minute capillaries in the skin. Papules and pustules (spots) accompany this disorder.
- ACNE VULGARIS – blocked and often infected sebaceous glands resulting in papules, pustules, comedones and in some cases, inflammation. Commonly affecting the face, neck, back and chest.
- AGNAIL – commonly known as a hangnail. The skin around the nail becomes loose and ragged.
- ALLERGIES – sensitivity causing the skin to react adversely by becoming red, hot and itchy.
- ALOPECIA – baldness caused by hair follicles being unable to produce new hairs. This may be in circular

areas such as on the scalp, known as ALOPECIA AREATA, or total hair loss from the scalp, which is known as ALOPECIA TOTALIS. Total loss of hair from the whole body is called ALOPECIA UNIVERSALIS.

- ASTEATOSIS – under activity of sebaceous glands causing excessively dry, scaly and often itchy skin.

- ATHLETES FOOT – technical term is TINEA PEDIS. A form of ringworm resulting from a fungal infection. It affects the soles of the feet and in between the toes and is characterised by itchy, soggy skin which flakes and peels.

- BARBER'S RASH – technical term is FOLLICULITIS. Affects areas of facial hair growth in men. Characterised by redness and swelling in the hair follicles.

- BED SORES – technical term is DECUBITUS ULCER. Painful, cracked, weeping areas of skin caused by irritation and continuous pressure on the body. Commonly found on buttocks, heels and elbows of bedridden people.

- BLISTER – a swelling caused by an accumulation of clear fluid under the skin. Commonly caused by injury to the skin, heat, irritant substances and friction.

- BOIL – bacteria attacks the hair follicle and surrounding skin. A pus formation develops which either comes to a head in the case of a boil or increases becoming an ABSCESS. A collection of boils is known as a CARBUNCLE.

- BRITTLE NAILS – also known as FRAGILITAS UNGUIUM. Excessively dry nails.

- BRUISE – discolouration in or below the skin tissue resulting from an injury. Blood leaks from damaged blood vessels into the surrounding tissue.

- CALLUS – thickening of skin caused by abnormal pressure or friction. Commonly found on the feet beneath a bone.

- CANCER – skin cancer is the most common type of cancer and is most often associated with exposure to ultraviolet radiation from the sun.

- CANITIES – grey or white hair due to lack of colour pigment forming in new hairs associated with ageing.

- CARBUNCLE – a collection of boils.

- CELLULITIS – spreading inflammation of skin. Skin is hot to touch, tender and shiny.

- CHERRY SPOTS – red or purple spots, which appear on the chest and trunk usually in later life. Harmless, bleed easily if injured but do not develop into cancer.
- CHILLBLAINS – painful, red, blue or purple areas of skin found on the toes and fingers as a result of poor circulation.
- CHLOASMA – small patches of dark colour, which appear as a result of melanocyte action being stimulated. Caused by a hormone imbalance and commonly associated with the contraceptive pill.
- COLD SORE – technical name is HERPES SIMPLEX. Caused by a virus, which can lie dormant in the sensory nerve cells for many years. It may be activated by stimuli such as fever, menstruation, extreme cold or sunburn. It may be primary (first attack) or secondary (recurring).
- COMBINATION SKIN – a combination of skin types on the face i.e. greasy centre panel and dry cheeks and neck.
- COMEDONE – technical term for a blackhead. The opening of a pore becomes blocked with excess dried sebum, dead skin cells, sweat and dirt etc. oxidising to form a blackened plug. More common in areas with more sebaceous glands i.e. centre panel of face.
- CORK SCREW HAIR – distortion of the hair follicle resulting in tightly curled hair.
- CORN – consists of a central core surrounded by thick layers of skin following repeated friction to an area – most commonly toes and soles of the feet.
- COUPEROSE SKIN – high colouring due to blood capillary network being close to the surface of the skin. Prone to damage resulting in broken capillaries.
- CYST – found in the sebaceous glands as a small painless lump. There are two kinds: epidermal cysts which appear anywhere on the body and are easily infected and the pilar cysts which usually appear on the scalp.
- DANDRUFF – technical term is PITYRIASIS CAPITIS and describes dry, flakey scalp.
- DECUBITUS ULCER – commonly known as BED SORES and characterised by painful, cracked and weeping skin on pressure points of the body.
- DEHYDRATED SKIN – skin lacking in moisture caused by inadequate fluid intake.
- DERMATITIS – another name for ECZEMA, an inflammation of the skin.

- DRY HAIR – hair that is lacking in moisture either because of under active sebaceous glands and/or use of harsh products and extreme temperatures.
- DRY SKIN – skin lacking in moisture due to under active sebaceous glands and or use of harsh products or extreme temperatures stripping the acid mantle. Dry skin is also a result of the ageing process.
- ECZEMA – also known as DERMATITIS. There are five types and are all forms of inflammation of the skin. Eczema starts with redness due to dilated blood vessels. Fluid accumulates in the skin causing swelling, itching and blisters. Weeping skin develops that may become infected. It eventually dries with scabs and crusts.
- EGGSHELL NAILS – very thin nails resulting from defective circulation to the germinal matrix.
- EMBEDDED HAIRS – hairs that do not emerge from the skin. The skin grows over the follicle trapping the hair below resulting in a small lump on the surface of the skin.
- EPHELIDES – see freckles.
- FISSURES – cracks within the epidermis, which exposes the dermis.
- FOLLICULITIS – affects areas of facial hair growth in men. It is also known as BARBER'S RASH and is characterised by redness and swelling in the hair follicles.
- FRAGILITAS UNGUIUM – the technical term used for BRITTLE NAILS.
- FRECKLES – technical term is EPHELIDES. Small, flat, irregular patches of melanin found on the face and the body. The melanin accumulates in small areas instead of being evenly distributed.
- FURROWS – ridges in the nail plate that may run transversely, these BEAU'S LINES are associated with temporary problems with the production of new cells within the germinal matrix, or longitudinally which are usually associated with age.
- GREASY HAIR – an over secretion of sebum coating the hair.
- GREASY SKIN – an over secretion of sebum causes an excess of moisture on the skin's surface. Associated with puberty but can occur due to general hormonal disturbances at any time of life.
- HANG NAILS – technical term is AGNAIL. The skin around the nail becomes loose and ragged.

- HERPES SIMPLEX – a viral infection commonly known as a cold sore. Most common site is the mouth.
- HIRSUTISM – abnormal pattern of hair growth involving a male pattern of hair growth in females.
- HIVES – technical term is URTICARIA. A swollen area of skin surrounded by a red wheal. The resulting eruptions resemble a nettle sting and are commonly known as nettle rash.
- HYPERIDROSIS – excessive perspiration affecting areas such as the hands, feet and underarms. Causes can be congenital and hormonal.
- HYPERTRICHOSIS – abnormal growth of hair involving hair growth in unusual areas of the body.
- IMPETIGO – highly contagious skin infection starting with a red spot which becomes a blister, quickly breaking down and discharging, with a yellow crust developing. The crust spreads – commonly affected areas include the face, hands and knees.
- INGROWING HAIRS – hairs which continue to grow below the surface of the skin. Prone to becoming infected unless loosened from the skin to grow normally.
- INGROWING NAILS – technical term is ONYCHOCRYPTOSIS. Nails grow into the nail wall piercing the skin. Prone to becoming infected and tend to affect the toes more than the fingers – in particular the big toe.
- KELOID – a harmless mass of excess tissue forming in the scar of an injury. More common in black skins.
- KERATOSIS, ACTINIC – hard, dry, scaly, flat brown growths that develop in later life. Most commonly found on exposed skin i.e. face and back of hands.
- KOILONYCHIA – or spoon-shaped nails. Commonly associated with iron deficiency, the nails grow in a spoon shape rising from the tips of the fingers or toes.
- LENTIGINES – a large, distinct freckle-like mark which may be slightly raised.
- LEUCONYCHIA – white spots on the nail plate caused by a superficial knock. These white spots grow out with the nail.
- MACULE – flat, coloured area of skin. May be lighter or darker than the surrounding skin.
- MATURE SKIN – skin that is ageing either prematurely because of the use of harsh products,

exposure to extreme temperatures, or naturally due to the passing of time and gradual decline of the skin's functions.

- MELANOMA – a pigmented mole. There are two types: harmless juvenile melanoma and malignant melanoma.
- MILIA – small, harmless pinhead cysts. Commonly found around the eyes and cheeks on dry skin types.
- MOLE – darkened, raised area of skin of varying size, shape and colour.
- MONILETHRIX – irregular development of hair in the follicle resulting in bead-like swellings along the hair shaft.
- NORMAL SKIN – well-balanced skin with all of its functions working in harmony.
- OILY HAIR – over secretion of sebum coating the hair.
- OILY SKIN – an over secretion of sebum causes an excess of moisture on the surface of the skin.
- ONYCHIA – infection of the nail fold resulting in inflammation.
- ONYCHOCRYPTOSIS – ingrowing nails commonly affecting the big toe.
- ONYCHOLYSIS – separation of the nail plate from the nail bed and is often associated with illness, overuse and abuse of false nails.
- ONYCHOPHAGY – nail biting often resulting in an exposed nail bed.
- ONYCHOPTOSIS – shedding of the nail plate, may be associated with ALOPECIA UNIVERSALIS if all nails are shed. If one nail is shed it may be caused by damage to the cells in the germinal matrix.
- ONYCHORRHEXIS – longitudinal splitting of the nail plate, associated with dry, brittle nails and ageing.
- ONYCHOSCHIZIA – flaking nails; the peeling of one or more layers of the nail plate due to a lack of moisture and fat.
- ONYCHOTILLOMANIA – involves the picking of the cuticle and/or nail resulting in ragged and often inflamed nails and cuticles with permanent damage possible if the germinal matrix is affected.
- ONYCHOTROPHIA – thinning nails that gradually waste away usually due to illness.
- PAPULE – a small raised area of unbroken skin. Solid and painful to touch often developing into a pustule.

- PARONYCHIA – inflammation of the skin surrounding the nail. May be caused by an infection or damage to this area.
- PIGMENTED NAEVI – harmless, coloured birthmark.
- PILI MULTIGENMINI – two or more hairs growing from the same follicle.
- PITYRIASIS CAPITIS – the technical term for DANDRUFF involving a dry, flakey scalp.
- PORT WINE STAIN – flat patch of dark red staining in the skin resulting from an abnormality of the capillary blood vessels. It is present at birth and is permanent.
- PRICKLY HEAT – technical term is MILIARIA RUBRA. A rash caused by excessive heat. Area of skin becomes red and itchy, and is accompanied by blocked or inflamed sweat glands.
- PSORIASIS – a recurring scaly eruption of the skin. Red patches develop covered by a scale which can be itchy.
- PTERYGIUM – overgrown cuticles which adhere to the nail plate.
- PUSTULE – often referred to as a white head, spot or pimple. Raised, pus-filled area of skin often developing from a papule.
- RINGWORM – a fungal infection which attacks the skin, known as TINEA CORPORIS, the nail, ONYCHOMYCOSIS and the feet, TINEA PEDIS.
- SCABIES – highly contagious disease caused by an infestation of a mite which burrows into the skin. Mainly found between the fingers, inside of the wrist and the soles of the feet. Characterised by itchy spots.
- SCAR – replacement tissue formed over the site of injury.
- SEBORRHOEA – over active sebaceous glands causing oily skin with enlarged, blocked pores, comedones and pustules.
- SENSITIVE SKIN – fine, pale skin whose protective functions are limited and need extra care.
- SHINGLES – technical term is HERPES ZOSTER. A blistered skin eruption over an area of skin by a nerve. A severely painful condition associated with later life or younger people who have had chicken pox.
- SKIN TAG – tiny, loose growths of skin. Colourless and painless, they develop on the neck, groin,

armpits and trunk at any age but most commonly in elderly women.

- SPIDER NAEVUS – small, painless, red spot with thin blood vessels leading from it making it look like a red spider. Commonly associated with pregnancy and can appear on the face, body and legs.

- STRAWBERRY NAEVUS – a bright red mark that may vary from a small pinhead to a swollen, irregular dome-shaped mark. Usually develops in the first week of life and is most commonly found on the head and neck.

- STRETCH MARK – thin lines of over-stretched skin. May be shiny and discoloured due to loss of elasticity in the dermis.

- SUNBURN – inflammation of the skin after excessive exposure to the sun's rays. Skin may blister as a result of further, unprotected exposure.

- SUPERFLUOUS HAIR – unwanted hair as a result of heredity, hormonal changes or stimulation to the area.

- TINEA PEDIS – a form of ringworm resulting from a fungal infection. It affects the soles of the feet and in between the toes, and is commonly known as ATHLETES FOOT.

- URTICARIA – technical term for HIVES and is characterised by a swollen area of skin surrounded by a red wheal.

- VITILIGO – white patches of skin caused by the destruction of melanocytes. It can sometimes be associated with hormonal imbalance.

- WARTS – small, solid growths on the skin. Caused by a virus and highly contagious. Also known as veruccas when found on the feet.

- WHEAL – a localised area of swelling.

- WHITLOW – an infection of the soft pad at the tip of the finger or thumb. Tender to touch with a pus formation.

System sorter

INTEGUMENTARY SYSTEM

Muscular

The condition and shape of the muscles which lie directly under the skin contribute to the general appearance of the skin. Poor muscle tone will be seen as dropped contours.

Respiratory

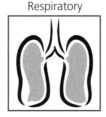

Oxygen present in the air enters the lungs as we breathe. It is taken to the cells of the skin, hair and nails via the blood to aid in their renewal.

Skeletal

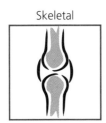

Vitamin D is needed for the formation and maintenance of bones. Vitamin D production in the skin is stimulated by the ultraviolet rays of the sun. Ergosterol is converted into vitamin D.

Circulatory

Blood clotting occurs at the site of an injury forming a scab on the surface of the skin. This protects the underlying structures from further damage as the skin heals.

Excess calories from the food that we eat are processed by the digestive system and stored as fatty (adipose) tissue in the hypodermis.

Digestive

Water is lost from the skin in the form of sweat. The kidneys help to regulate fluid balance to help prevent the skin from becoming dehydrated.

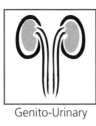

Genito-Urinary

Sensory nerves link the skin to the brain and provide an 'early warning' system against external attack.

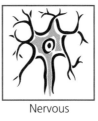

Nervous

Hormones produced by the endocrine glands regulate the activity within the integumentary system eg hair growth is affected by sex hormones at times like puberty, pregnancy and menopause.

Endocrine

The skin, hair and nails contribute to the general well-being of the body by providing an external covering that acts like a suit of armour keeping vital systems IN and harmful invaders OUT! The survival of the integumentary system relies on the inter-balance between all of the other systems of the body.

Holistic harmony

The integumentary system relies on the internal and external forces of nature to keep it balanced. The internal functions of the human body together with the right type of external environment will allow the skin, hair and nails to function well and provide us with a covering that is both protective in its function and attractive in its appearance.

There are many factors that need to be considered in this battle for balance including:

Fluid

The skin, hair and nails will benefit from us drinking plenty of water a day. Drinking water should ideally be still and at room temperature to aid digestion (Chapter 7). Carbonated water introduces too much air into the stomach whereas still water has a much gentler effect. In addition, herbal teas aid the hydration of the skin, hair and nails; they also have specific actions on other systems that in turn have a positive effect on the integumentary system:

- Ginger tea aids circulation, which in turn improves skin colour.
- Peppermint tea aids digestion, which in turn improves the value of nutrients taken to the integumentary system helping cell reproduction.
- Chamomile tea aids relaxation, which in turn has a calming and relaxing effect on the body and the system as a whole.

Nutrition

The food we eat will be digested by the digestive system (Chapter 7) and taken to the cells of the integumentary system via the blood (Chapter 6) for cellular metabolism to take place in the form of growth, development, renewal and repair. The evidence of this is clearly seen on the surface of the skin, hair and nails – we are what we eat!

Examples of foods with special value for this system include:

- Vitamin C in the form of citrus fruits for collagen development and maintenance in the dermis.
- Vitamin A – Retinol found in meat/fish/dairy products and Betacarotene found in yellow, red and orange fruits and vegetables to help prevent rough, dry skin and dandruff.
- Zinc found in oysters, nuts, peas and beans to aid skin elasticity and help prevent stretch marks, as well as to keep hair and nails healthy.

- Vitamin B2 in the form of mushrooms, watercress and cabbage to help to repair and maintain healthy skin, hair and nails.
- Vitamin E in the form of seeds and nuts to assist healing.

A varied, balanced diet rich in fresh vegetables and fruit will contribute to clear skin, shiny hair and healthy nails. When we eat exactly what we need in the way of fuel, there is no wasted 'burning' by the body therefore less free radical activity takes place.

Rest

Surface skin cells are constantly being shed (desquamated) from the body and are replaced by new cells, hair follows a growth cycle that involves the production of new cells by mitosis, and nails are also reliant on the constant renewal of cells. This process is greatly improved when the body is resting. On average, adults need 6–8 hours of quality sleep in every 24 hours. This rest time allows the body, and so the skin and its parts, to replenish and regenerate efficiently. Without adequate resting time, the skin, hair and nails become dull and lifeless with the finer skin around the eyes developing dark shadows. In addition, the skin appears to age dramatically when the body is tired and in need of rest. It is important, therefore, to listen to the signals that the body gives out and even more important, to respond to them! The skin, in turn, is quick to respond to adequate rest and will soon show the positive results of improved activity within the cells.

Activity

In contrast, physical activity stimulates the circulation ensuring an active blood supply to the skin and its parts as well as to other parts of the body (Chapter 6). This in turn improves skin colour, hair texture and the appearance of the nails as well as overall efficiency. Poor circulation caused by inactivity results in cold, pale, skin and dull hair and nails. As the skin covers the whole body, it is important that the whole body takes part in some form of activity in order to stimulate it. Brisk walking and swimming are particularly good examples of physical activities that have an overall stimulating effect on the system, improving cellular function (respiration and metabolism) in the hypodermis, dermis and epidermis. Massage also has a similar effect on the system stimulating the necessary functions so that the appearance, efficiency and effectiveness are greatly improved.

Air

The quality of the air we breathe has a direct effect on this system:

- Toxins from the environment are absorbed into the skin breaking down the acid mantle and its protective functions resulting in vulnerable, problem skin, hair and nails.

- Air conditioning and central heating both have a drying, dehydrating effect on the skin, hair and nails, stripping their surfaces of necessary moisture.

- Extreme weather conditions affect the skin, hair and nails dramatically resulting in dehydration and damage that requires additional protection and care.

- Smoking is one the single most harmful factors to affect the skin and its parts. Passive smokers absorb the toxins from the smoke into the various layers of the skin, hair and nails, making them smell, changing their colour and affecting their vital functions. Respiration is affected as smoke is inhaled and taken to the cells via the interchange of air into the blood within the lungs. Active smokers suffer from all of the aforementioned, but at more concentrated levels, as well as suffering from the added problem of the direct drying effect of the smoke on the skin around the eyes and lips. This contributes greatly to premature ageing of the skin together with the associated problems of lines and wrinkles. Therefore, not only should thought be given to the methods of breathing for holistic harmony, but also to the environment and quality of the air we are breathing in. Fresh, unpolluted air combined with regular deep breathing exercises ensures the body is energised instead of de-energised!

Age

A factor which is out of our direct control, and unfortunately for the skin, hair and nails, can have a seriously debilitating effect on their structure, function and thus, appearance. As the body ages, its systems slow down and become less active, the results of which become apparent on the skin's surface. The collagen, elastin and reticulin fibres in the dermis gradually start to break down and the skin loses the youthfulness and firmness that has become so desirable. The skin and hair also begin to lose their flexibility and lustre, gradually becoming thinner and finer with age.

- A young skin = an inexperienced body and mind!
- A mature skin = a body and mind enriched with life's experiences!

There is always the possibility of balance – the harmony is in its acceptance!

Sebaceous and sudoriferous glands in the dermis become less active and the skin loses its suppleness as lines and wrinkles begin to develop. Hair and nails become dry and brittle. Melanin production changes producing white hair and brown age spots! The skin and its parts become more prone to problems as its protective functions lessen, and natural changes occur that we sometimes find difficult to come to terms with. Although the passing of time is out of our control, the quality of time is very much under our direct control. Therefore, we need to take time out to care for our skin and its parts, and to become more aware of their changing needs as time passes, ensuring that those needs are being met, and most importantly, remembering to be thankful for what we have got!

Colour

Colour can be used in many ways to benefit the integumentary system. Each colour vibrates at its own frequency, as does everything else including the human body. When the body is healthy the frequency of these vibrations is balanced. However, the effects of stress and illness can disturb the balance and the use of colour can help to restore it. This can be done by visualisation of colour or by actually physically working with a colour and can be as simple as using blue towels to induce a feeling of calm compared with red towels which would have the opposite effect. As the integumentary system provides a first point of contact for the body, it is instantly affected by the colours that surround it. On a basic level colour will have the following effect on the system:

- Red and orange help to stimulate.
- Yellow helps to refresh.
- Green helps to de-stress.
- Blue helps to calm and soothe.
- Violet helps to rejuvenate.

Awareness

The skin can be classified as different types: *normal, oily, dry, sensitive, dehydrated, mature* and *combination*.

- Normal skin is usually associated with young skin up until puberty. The skin has perfect balance and is clear, fresh and healthy.
- Oily skin occurs as a result of over activity in the sebaceous glands. Excess sebum is produced, often

during puberty causing blocked pores, comedones (blackheads) and papules and pustules (spots).

- Dry skin occurs as the sebaceous activity within the skin slows down. The lack of sebum causes the skin to feel tight and sometimes sore as the protective functions are diminished.

- Sensitive skin is usually fine, pale and easily stimulated. The blood capillary network is close to the surface and gives sensitive skin its characteristic redness.

- Dehydrated skin suffers from a lack of moisture as a result of illness, lack of water in the diet, excessive sun bathing, harsh products etc. The sebum evaporates from the surface of the skin leaving the skin dry, tight and vulnerable.

- Mature skin occurs with the ageing process as sebaceous secretion diminishes and skin and muscle tone become more relaxed. There is a general deterioration in the collagen, elastin and reticulin fibres of the dermis, and the skin begins to sag and wrinkle. Premature ageing may occur as a result of the factors associated with dehydrated skin.

- Combination skin occurs when there is a combination of skin types present at the same time e.g. the most common combination involves an oily center panel with normal cheeks and neck.

An awareness of the different skin, hair and nail types forms the basis of treatment and product recommendations.

Special care

The general physical and psychological well-being of every living being is dependent on some form of loving care. Not only is it necessary to enjoy the love of others – a loving word, deed or touch heals like a balm, soothing body and mind, somehow having the ability to right all wrongs. It is also necessary to apply this philosophy to care for ourselves – a cared for body and mind leads to cared for skin just as cared for skin leads to a cared for body and mind. Positivity contributes to successful loving care resulting in the feeling of confidence and raised self-esteem together with increased physical well-being in much the same way as negativity contributes to feelings of despair, lethargy and lack of self-worth. These feelings are reflected in the general condition of the body and mind and can be 'seen' through the skin and its parts, providing a physical showcase for our innermost feelings e.g. the glow associated with first love, the radiance associated with pregnancy. This is sometimes called the 'look of love'. Nails clearly show the condition of the blood circulation. If this is good down to the extremities i.e. the fingers and toes, it reflects a healthy body.

Our skin has the ability to provide a mirror image of our levels of well-being and the effects of the loving care or abuse given to the body. It highlights for us the need to have an awareness of the importance of this type of care not only in the short term to aid a specific problem, but also in the long term, in order to maintain holistic harmony.

The skin, hair and nails also rely on external care to maintain their structure, function and appearance, in the form of:

Angel advice

Vitamins A, C and E may be added to these products to help to combat free radical activity.

- Cleansing – all external parts of this system need to be kept clean with the use of:
  - Facial cleansers suitable for specific skin types
  - Shampoos for specific hair types
  - Soaps for hands, nails and body.
- Exfoliating – is necessary to aid the natural desquamation process of the skin.
- Granules – e.g. wheat germ, will have an abrasive effect on the skin.
- **AHA's** – **a**lpha **h**ydroxy **a**cids or *fruit acids* e.g. citric acid produces a stimulating effect on cellular turnover providing a natural exfoliation.
- Toning – helps to tighten pores keeping the right amount of moisture within the skin.
- Masks – aid specific skin, hair and nail problems e.g.:
  - Clay masks for excessively oily skin
  - Nourishing conditioning masks for dry hair
  - Hot oil treatment for flaky nails.
- Moisturising – will replace lost moisture from the external structures of the system in the form of:
  - Day/night creams for face and neck
  - Remedial creams/gels for eyes
  - Lotions/creams/oils for general body
  - Creams for hands and nails
  - Conditioners for hair.

The integumentary system therefore relies on the balance between the internal and external care of the skin, hair and nails to maintain holistic harmony.

Treatment tracker

INTEGUMENTARY SYSTEM

Make up

Facials

Nail care

The use of make up can enhance or detract depending on the needs of the skin. Make up can be used to enhance a good skin and can be used to cover the results of a bad skin. Hair colour and nail enamel are a form of make up.

The procedures used in a facial have a direct effect on the skin with beneficial results. Choice of products may be adapted to aid all skin types and skin problems.

Hair removal

Nail care procedures stimulate the growth of the nails by increasing blood flow to the area. This in turn improves the efficiency of cell reproduction (mitosis) encouraging the appearance of a healthy nail plate and free edge

Depilation removes excess hair temporarily. Epilation removes hair permanently by destroying the hair bulb until it is no longer able to reproduce a new hair.

Massage has a beneficial effect on the integumentary system as a whole. It can be an invigorating treatment stimulating the growth of healthy skin, hair and nails. The introduction of soothing, nourishing products can be an additional aid

Massage

The skin, hair and nails benefit from a gentle reflexology treatment where all the reflex points have been worked. Body systems that are balanced and working in harmony with one another result in the appearance of healthy hair, nails and skin.

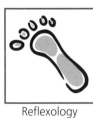

Reflexology

Essential oils can be blended to suit specific problems associated with the integumentary system eg dry skin, hair and nails will benefit from a blend of jasmine, lavender and chamomile (R).

Aromatherapy

Galvanic face and body treatments have a beneficial effect on the skin in the short and long term. In the short term the skin gains benefit from desquamation of dead cells and absorption of products. In the longer term, the skin benefits from the stimulation of blood flow, speeding up mitosis in the stratum germinativum.

Electrical

The integumentary system benefits from the results of beauty and holistic treatments on both a superficial and a deep level. The superficial appearance of the skin is improved as a direct result of treatments and the stimulation of the deeper structures provides more long term benefit with improved cellular functions.

Knowledge review – Integumentary system

1   How many main layers make up the skin, the hair and the nails?

2   What is melanin and where is it found?

3   What is sebum and where is it produced?

4   What are the benefits of sebum for the skin and hair?

5   What is mitosis and in which three areas of this system does it take place?

6   Describe the fibres that make up the dermis or true skin.

7   What is the acid mantle made up of and what are its functions?

8   How does the digestive system contribute to the well-being of the skin/hair/nails?

9   How long does it take for the cells of the epidermis to reach the skin's surface?

10   How does the skin react to the sun's rays?

11   Name and describe the different stages of hair growth.

12   Which layer of the epidermis contributes to the waterproofing of the skin?

13   Give three effects of ageing on the skin/hair/nails.

14   Which body system supplies the cells with the ability to feel and respond?

15   How do the skin/hair/nails benefit from fluid?

16   How does the system respond if the body is suffering from lack of sleep?

17   Give three ways in which the skin and its parts protect the rest of the body.

18   How does the skin assist in temperature control?

19   What are the structures called that are responsible for 'goose bumps'?

20   Why is smoking harmful for the parts of the integumentary system?

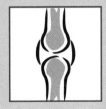

The skeletal system

3

Learning objectives

After reading this chapter you should be able to:

- **Recognise the position and structure of the bones of the skeleton**

- **Identify the bones and main joints of the body**

- **Understand the functions of the skeletal system**

- **Be aware of the factors that affect the well-being of the skeletal system**

- **Appreciate the ways in which the skeletal system works with the other systems of the body to maintain homeostasis.**

The next part of our journey through the human body takes us below the skin and muscles to the system that provides us with our basic shape and structure, **THE SKELETAL SYSTEM**, consisting of 206 individual bones together with the formation of joints. This system provides the body with a framework that is solid in structure and flexible in function. As we explore the anatomy and physiology of this system further, we will begin to appreciate its supportive and protective functions as well as the links it maintains with all of the other systems of the body.

Science scene

Structure and position of bones and joints

The skeletal system is made up of dense connective tissue, which forms the **bones** of the body. Bones are the hard structures which make up the basic framework of the skeleton. Where two or more bones meet a **joint** is formed.

The skeletal system also comprises of tough connective tissue forming supportive structures in the form of **cartilage**, **ligaments** and **tendons**.

- Cartilage provides the skeletal system with connection, support, flexibility and protection.
- Ligaments connect bones at joints allowing movement between two or more bones.
- Tendons attach muscle to bone.

Bones

Bones are the most rigid structures of connective tissue and differ greatly in size and shape but are similar in structure, development and functions.

Bones are made up of active living tissue containing:

- Water – approximately 25 per cent
- Inorganic material – approximately 45 per cent comprised of calcium and phosphorus
- Organic material – approximately 30 per cent comprised of bone forming cells called **osteoblasts** together with a blood and nerve supply.

Bone formation

Because they are living tissue, bones grow during childhood, will bleed and hurt if damaged and are able to repair themselves when broken. A hardening process called **ossification** takes place as bones mature making them solid structures that can withstand a great deal. In addition bones contain collagen, which provides them with resilience, and calcium, which gives them strength. Many bones contain an inner cavity containing **bone marrow**. Bone marrow is either red or yellow. Red bone marrow produces new blood cells and yellow bone marrow is a storage area for fat cells.

Like the epidermis of the skin, bones are continuously renewing themselves, but unlike the epidermis, bones do

Fascinating Fact

Fossilised bones provide scientists with a glimpse of our past

!

Remember

Excess fat is stored in the hypodermis below the skin and in the yellow marrow of the bones. These fat stores can be used for energy production in the absence of food!

this very slowly. Special cells called **osteoclasts** break down old bone cells and osteoblasts then form new bone cells. As bone cells mature they become known as **osteocytes.**

There are two types of bone tissue **compact** or hard bone tissue and **cancellous** or spongy bone tissue.

Compact bone tissue

Compact bone tissue has an almost solid inner structure making it hard and durable.

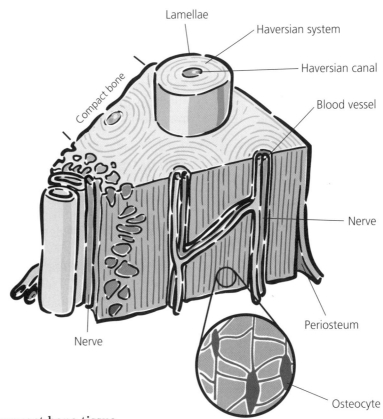

Compact bone tissue

It consists of several **haversian systems** that each contain:

- A central *haversian canal* containing blood and lymph vessels, and nerves for 'feeding' (cellular respiration and metabolism) and 'feeling' (sensation).
- Plates of bone called **lamellae**, which are placed around the haversian canal forming a solid structure which has great strength.

Cancellous bone tissue

Cancellous bone tissue has a looser inner structure giving a spongy appearance to the bone. It has much larger haversian canals and less lamellae, providing spaces

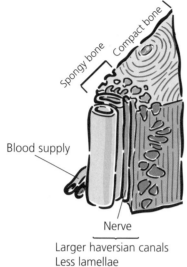

Blood supply

Nerve

Larger haversian canals
Less lamellae
Spaces between haversian systems

Cancellous bone tissue

Cartilage does not repair or renew itself as easily as bone. Bones receive a richer supply of blood than cartilage and it is this blood supply that provides the nutrients for growth and repair.

between the **haversian systems**, which are filled with the red or yellow bone marrow.

All bones are made up of a varying combination of compact and cancellous tissue depending on their size, shape and function.

Bones also have an outer covering of **periosteum** or **cartilage** giving them extra protection, strength and durability.

- Periosteum covers the length of bones.
- Cartilage covers the ends of bones at a joint.

Periosteum

Periosteum is made up of two layers: the inner most layer is responsible for producing new cells for the growth and repair of bone tissue and the outer layer contains a rich blood supply enabling bones to receive the nourishment needed for growth and repair.

Cartilage

Cartilage is comprised of firm connective tissue made up of elastin and collagen fibres, making it flexible and durable. There are three types of cartilage:

1. *Hyaline cartilage*, sometimes called *articular* cartilage, covers the ends of bones where they meet at a joint and helps to prevent friction as bones move against one another. It also helps in the attachment of some bones to others e.g. the ribs to the breastplate, and it forms parts of the nose and windpipe.

2. *Fibrocartilage* is less flexible but slightly tougher and forms pads between the bones e.g. the spine.

3. *Elastic cartilage* is very flexible and forms the areas of the body that need to move more freely e.g. the ear.

Ligaments

Ligaments are formed from fibrocartilage as tough connective tissue that links bones together at a joint. They allow the bones to move freely within a safe range of movement. The ligaments pull tightly to prevent any movement that will cause damage to the bones.

Tendons

Tendons are made up of striped bundles of collagen fibres which attach muscle to bone e.g. the Achilles tendon attaches the calf muscle to the foot at the ankle. When the tendon is broad and flat, like those attaching the muscles of the head to the bones of the skull, it is known as an **aponeurosis**.

Types of bones

The skeleton is made up of different types of bones with various positions and function within the body. There are five different types of bones: *long*, *short*, *irregular*, *flat* and *sesemoid*:

1. Long bones are limb forming i.e. arms and legs. They are longer than they are wide.

2. Short bones are small. They are as broad as they are long and are round or cube-shaped e.g. the bones of the wrist.

3. Irregular bones are different shapes and sizes e.g. the bones of the spine.

4. Flat bones are thin, flattened in shape and generally curved e.g. the shoulder blades.

5. Sesemoid are small bones located within tendons e.g. the kneecap.

The skeleton and types of bone

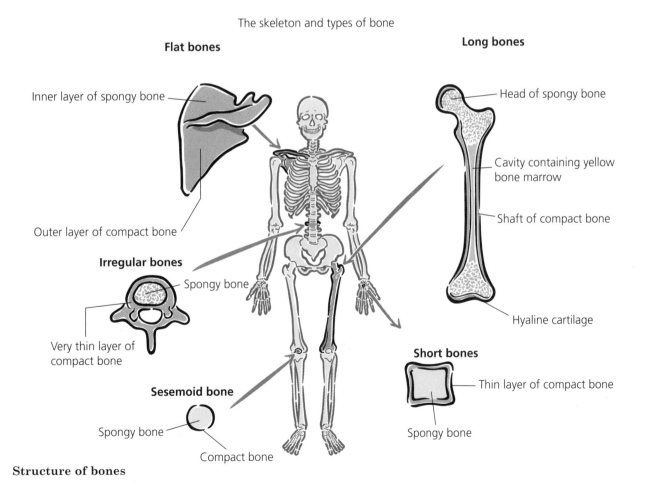

Flat bones

Long bones

Inner layer of spongy bone

Head of spongy bone

Cavity containing yellow bone marrow

Shaft of compact bone

Outer layer of compact bone

Irregular bones

Spongy bone

Very thin layer of compact bone

Hyaline cartilage

Short bones

Thin layer of compact bone

Sesemoid bone

Spongy bone

Spongy bone

Compact bone

Structure of bones

Remember that yellow bone marrow stores fat cells in the bones and red bone marrow is responsible for forming new blood cells.

The length of long bones is made up mainly of **compact** bone tissue, which in long bones, contains an inner cavity filled with yellow bone marrow. The ends are made up of cancellous bone tissue which contains red bone marrow.

Short, irregular, flat and sesemoid bones are made up of cancellous bone tissue containing red bone marrow surrounded by compact bone tissue containing no bone marrow. Some bones, for example facial bones, also have air spaces called **sinuses**, which make them lighter.

Bone development

The development of the skeletal system is continuous throughout life with bones reaching a certain thickness, length and shape when a person has reached the age of twenty-five. Bones continue to develop with the death of old bone cells and the production of new bone cells maintaining their structure. Factors that affect the development of bones include:

- **Genetic** – individual characteristics such as length and thickness of bones are inherited.
- **Nutrition** – a balanced diet high in vitamin D and minerals such as calcium maintains healthy bone development. Vitamin D enables the calcium to be absorbed from the digestive system where it is taken by the blood to the bones to be stored. This storage of calcium contributes to the solid structure of bones.
- **Hormones** – affect the growth and development of bones. Hormones are chemical messengers which are sent via the blood supply to the bones. They are responsible for informing the bones about when to stop growing etc.

The skeletal system is able to repair itself when damaged. When a bone is fractured (broken) the following processes take place:

1. A blood clot forms at the site of the break.

2. **Osteoblasts** form new bone tissue.

3. **Osteoclasts** 'mop up' old bone cells and ensure that the new bone tissue develops in the right shape.

This process can be helped by the use of casts, metal plates, screws etc. to keep the bone immobilized whilst it heals.

Fascinating Fact

Some synovial joints contain a bursa – a sac-like extension of the synovial membrane which provides additional cushioning where tendons rub against bones or other tendons, preventing excessive friction and damage.

Joints

A site where two or more bones come together is called a joint. There are three different types of joints which are classified according to the amount of movement they allow and include **fibrous**, **cartilaginous** and **synovial** joints:

1. Fibrous joints are also known as **sutures** and are joints that allow *no* movement between bones e.g. skull in an adult. They are sometimes classified as *fixed joints* because of this lack of movement.

2. Cartilaginous joints contain fibrocartilage pads between the bones. These joints allow limited movement which is created by the bones pushing down on the pads of cartilage. These are also classed as *slightly movable joints* e.g. joints of the spine.

3. Synovial joints contain hyaline cartilage at the ends of the bones where they meet and are held together by ligaments. The cavity between the bones contains a substance called **synovial fluid** which allows ease of movement in the joint. For this reason, synovial joints are also known as *freely moveable joints*.

Classification of joints

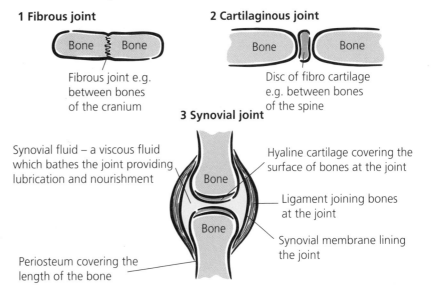

1 Fibrous joint

Fibrous joint e.g. between bones of the cranium

2 Cartilaginous joint

Disc of fibro cartilage e.g. between bones of the spine

3 Synovial joint

Synovial fluid – a viscous fluid which bathes the joint providing lubrication and nourishment

Hyaline cartilage covering the surface of bones at the joint

Ligament joining bones at the joint

Synovial membrane lining the joint

Periosteum covering the length of the bone

Fascinating Fact

The 'cracking' of joints e.g. knuckles is associated with a bubble forming in the synovial fluid due to a change in pressure. When this bubble bursts the distinctive 'crack' can be heard.

Synovial joints are of special interest to a therapist because our massage treatments help to keep them freely movable by warming the fluid surrounding the joint which allows greater ease of movement.

Synovial joints

There are seven different types of synovial joints: **plane**, **pivot**, **hinge**, **ellipsoid**, **condyloid**, **saddle** and **ball and socket**.

Types of synovial joints

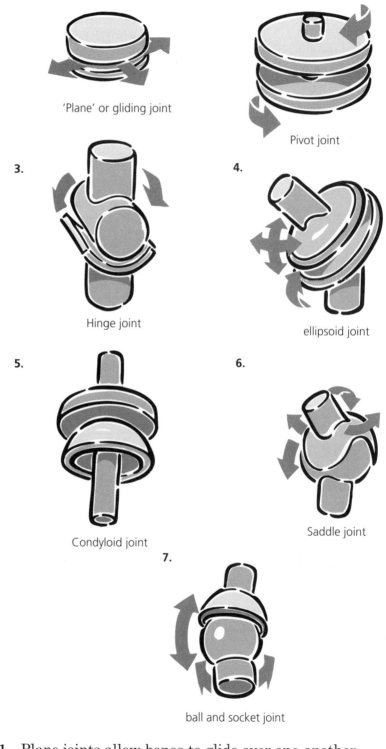

1.

'Plane' or gliding joint

2.

Pivot joint

3.

Hinge joint

4.

ellipsoid joint

5.

Condyloid joint

6.

Saddle joint

7.

ball and socket joint

1. Plane joints allow bones to glide over one another.

2. Pivot joints allow a circular movement known as *rotation*.

3. Hinge joints allow movements of flexion and extension i.e. bend and straighten.

4. Ellipsoid joints allow movements of flexion, extension, abduction (away from the body) and adduction (towards the body).

5. Condyloid joints allow movements of flexion and extension, abduction and adduction and a small amount of rotation.

6. Saddle joints are similar to condyloid joints but allow more rotational movement.

7. Ball and socket joints allow all movements in all directions.

The skeleton

Now that we have covered the structures that make up the skeletal system and understand the links between the

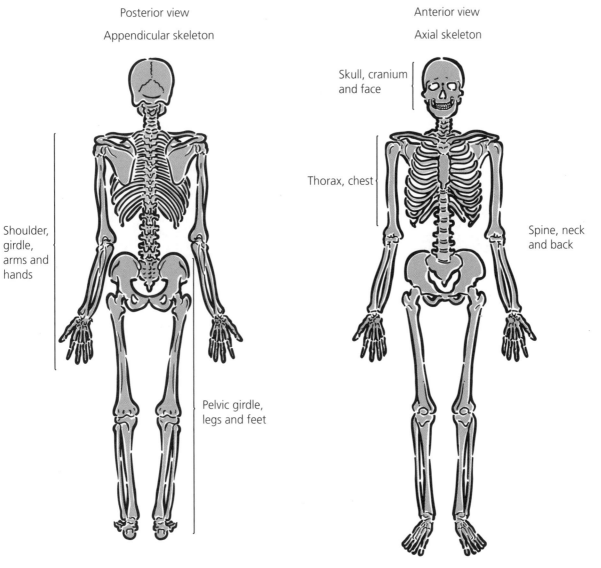

Posterior view

Appendicular skeleton

Anterior view

Axial skeleton

Skull, cranium and face

Thorax, chest

Shoulder, girdle, arms and hands

Spine, neck and back

Pelvic girdle, legs and feet

Skeleton – posterior and anterior views

bones, cartilage, ligaments and tendons, we can look at the skeleton as a whole. We need to learn to identify the bones and joints that make up the human skeleton in order to appreciate the way in which the body is able to support itself and move.

The human skeleton is made up of two parts, the *axial* and *appendicular* skeleton. The axial skeleton is comprised of the:

- **Skull** – the cranium and the face
- **Spine** – the neck and back
- **Thorax** – the chest.

The appendicular skeleton is comprised of the:

- Shoulder girdle arms and hands
- Pelvic girdle legs and feet.

The skull

The skull is made up of bones forming the **cranium** and the face most of which are flat or irregular and have sutures. Their main function is to protect the brain.

The cranium is comprised of eight bones:

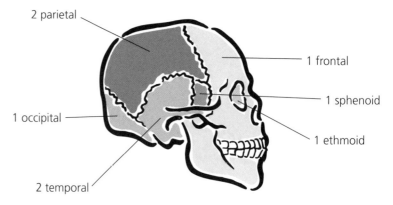

Bones of the cranium

- 1 **frontal** bone forms the *forehead* and contains two sinuses, one above each eye.
- 2 **parietal** bones form the *crown* of the skull.
- 1 **occipital** bone forms the *base* of the skull and contains an opening for the spinal cord linking the brain with the rest of the body.
- 2 **temporal** bones form the *temples* at the sides of the skull.
- 1 **ethmoid** bone forms part of the *nasal cavity* and contains many small **sinuses** at the sides of each eye.

Fascinating Fact

The sinuses are air spaces in the bones which are lined with epithelial tissue containing goblet cells that secrete mucus. They are connected to the nasal cavity by passageways. These sinuses become blocked up when we have a cold and massage can help to drain them. The other functions of these sinuses are to reduce the weight of the skull and increase the intensity of the voice as they act as sound chambers.

- 1 **sphenoid** bone forms the *eye socket* and contains two sinuses, one either side of the nose.

The face is comprised of 14 bones:

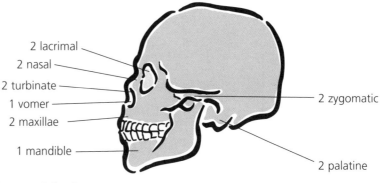

2 lacrimal
2 nasal
2 turbinate
1 vomer
2 maxillae
1 mandible
2 zygomatic
2 palatine

Bones of the face

Fascinating Fact

There is an additional, isolated bone called the **hyoid** bone lying at the front of the neck forming a base to which the tongue attaches. This is a sesemoid bone.

- 2 **zygomatic** bones form the *cheeks*.
- 2 **maxillae** bones fuse together to form the *upper jaw* containing the sockets for the upper teeth as well as the largest pair of sinuses.
- 1 **mandible** bone forms the *lower jaw* containing sockets for the lower teeth. It is the only movable bone of the *skull* forming a synovial ellipsoid joint with the temporal bone allowing us to move the mouth when we talk and chew!
- 2 **nasal** bones form the *bridge* of the *nose*.
- 2 **palatine** bones form the *floor* and *wall* of the *nose* and *roof* of the *mouth*.
- 2 **turbinate** bones form the *sides* of the *nose*.
- 1 **vomer** bone forms the *top* of the *nose*.
- 2 **lacrimal** bones form the *eye sockets* containing an opening for the *tear duct*.

The spine

The spine or *vertebral column* is made up individual bones or **vertebrae** which are irregular in shape and have cartilaginous joints except for the first two vertebrae, which form synovial joints. The spine provides protection for the spinal cord and can be divided into five sections:

1. **Cervical** – contains seven bones forming the length of the neck and the top of the back. The first bone, the *atlas* vertebra, supports the skull at an ellipsoid joint with the occipital bone, and the second bone, the *axis* vertebra, allows the rotation movement of the head through a pivot joint with the atlas bone.

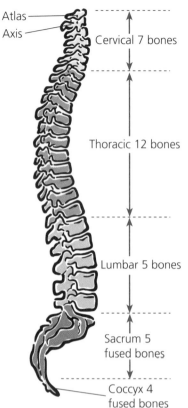

Atlas
Axis
Cervical 7 bones

Thoracic 12 bones

Lumbar 5 bones

Sacrum 5
fused bones

Coccyx 4
fused bones

The spine

2. **Thoracic** – contains 12 bones forming the upper and mid back to which 12 pairs of ribs attach.

3. **Lumbar** – contains five bones forming the lower back.

4. **Sacrum** – contains five fused bones forming the base of the spine.

5. **Coccyx** – contains four fused bones forming the tail.

The thorax

The thorax is made up of flat bones and forms the *chest*. It provides a protective cavity for the heart and lungs. The bones and synovial joints that make up the thorax include:

● 12 **thoracic vertebrae** of the spine at the back of the body.

● 12 pairs of **ribs** forming a cage like structure at the front of the body.

There are plane joints between the ribs and the vertebrae allowing a slight gliding movement enabling the chest to expand when we breathe in.

● Each of the ribs attach at the back to a thoracic vertebra.

● The first seven pairs of ribs attach at the front to the **sternum** (the breast plate) and are called the true ribs.

● The next three pairs of ribs attach to the ribs above them and are called false ribs.

● Finally, there are 2 pairs of ribs that do not attach to anything at the front and are known as floating ribs.

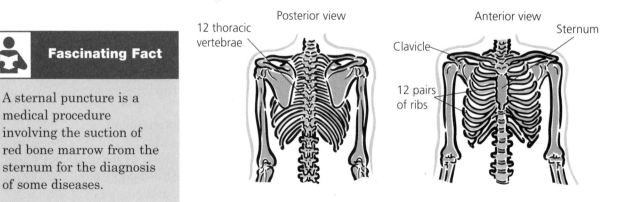

Posterior view
12 thoracic vertebrae

Anterior view
Sternum
Clavicle
12 pairs of ribs

The thorax

The shoulder girdle and arms

The shoulder girdle and arms are made up of the following bones and synovial joints:

- **Scapulae** form the *shoulder blades* and are flat bones.
- **Clavicles** form the *collarbones* and are long bones.

The joint between these bones is a plane allowing a slight gliding movement.

- **Humerus** bones form the upper arms and are long bones.

The scapula and humerus bones form the ball and socket shoulder joint allowing a full range of movement.

- **Ulna** and **radius** bones form the forearm and are long bones.

The synovial joint at the elbow between the three bones of the arm is a hinge allowing flexion and extension. The joint between the ulna and radius at the elbow is a pivot allowing rotational movements as well. These rotation movements allow the arm to **supinate** – turn so that the

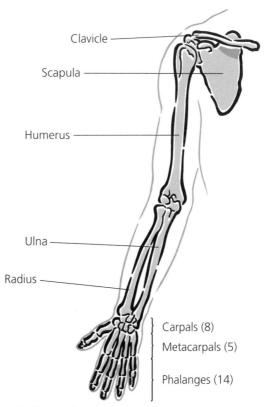

Clavicle

Scapula

Humerus

Ulna

Radius

Carpals (8)

Metacarpals (5)

Phalanges (14)

The shoulder girdle and arms

palm of the hand is facing upwards, and **pronate** – turn so that the palm of the hand is facing downwards.

- 8 **carpal** bones form each *wrist* and are short bones.

The joint at the wrist between the radius and the carpals is an ellipsoid allowing movements of flexion, extension, abduction and adduction. Plane joints exist between the individual carpal bones offering a gliding movement.

- Five **metacarpal** bones form the *hand* and are miniature long bones.
- Three **phalange** bones make up each *finger* and are miniature long bones.
- Two phalange bones make up each *thumb* and are miniature long bones.

There are a total of 14 phalange bones to each hand.

The thumb forms a saddle joint with the carpal and metacarpal bones allowing an almost full range of movement. The fingers form condyloid joints where the phalange bones meet the metacarpal bones allowing slightly less movement, and the joints between the bones of the fingers and thumbs are all hinges, allowing flexion and extension only.

The pelvic girdle and legs

The pelvic girdle and legs are made up of the following bones and synovial joints:

- Sacrum and coccyx forming the base of the spine and the middle of the pelvis.
- **Coxae** forming the protruding *hip* bones attaching to the sacrum and coccyx by fibrous joints.
- Each coxa is made up of three fused flat bones:
  1. **Ilium**, i.e. *groin*.
  2. **Pubis**, i.e. *pubic area*.
  3. **Ischium**, i.e. hip.
- **Femur** bones make up the *thigh* and are long bones.

The large joint between the coxa and femur bones is a ball and socket allowing a full range of movement.

- **Tibia** and **fibula** bones make up the *shin* and are long bones.

Activity

While reading this, move your own joints to check their range of movements.

Remember

The circular gap in the middle of the pelvic girdle makes up part of the birth canal in females.

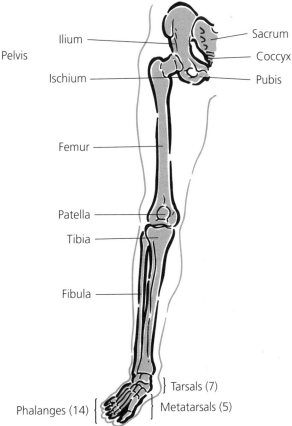

Ilium — Pelvis — Ischium — Sacrum — Coccyx — Pubis — Femur — Patella — Tibia — Fibula — Tarsals (7) — Metatarsals (5) — Phalanges (14)

The pelvic girdle

Tip

It is worth noting that books may give differing classifications of joints due to the slight differences in action within ellipsoid, condyloid and saddle joints. This can be confusing and annoying but the main focus should be on the fact that they are all movable joints that share a similar range of movement.

Fascinating Fact

Injuries to elbow, shoulder and knee joints are often diagnosed by a thin tubular instrument called an **arthroscope,** inserted into the joint capsule. Optical fibres transmit the image onto a video screen.

- **Patella** bones form the *knee cap* and are sesemoid bones.

The joint between the femur and tibia at the knee is a hinge allowing the knee to bend and straighten.

- Seven **tarsal** bones make up the *ankle* and are short bones.

The joint at the ankle between the tibia, fibula and a tarsal is an ellipsoid allowing the foot to bend, stretch and move inwards and outwards. These four movements are called:

1. **Dorsi flexion** – bending the foot upwards
2. **Plantar flexion** – stretching the foot downwards
3. **Eversion** – turning the foot out
4. **Inversion** – turning the foot in.
   - Five **metatarsal** bones make up the *foot* and are miniature long bones.
   - Three **phalange** bones make up each *small toe* and are miniature long bones.

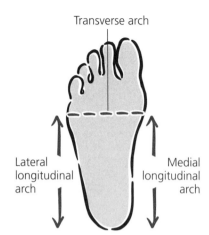

Transverse arch

Lateral longitudinal arch

Medial longitudinal arch

The arches of the feet

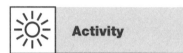

Activity

Take some time to analyse your own skeletal system, locate and feel your various bones, move your joints and check out your arches. You will be amazed at their structure and functions and it will make you very conscious of how we take our body for granted!

Remember

It is important to be aware of the fact that the skull provides protection for 12 pairs of cranial nerves coming from the brain and the spine provides protection for 31 pairs of spinal nerves coming from the spinal cord. These nerves form part of the peripheral nervous system (Chapter 9) that allows the face and body to receive and relay messages. Damage to the skull or spine can affect the healthy working of these nerves resulting in loss of function i.e. paralysis.

- Two **phalange** bones make up each *big toe* and are miniature long bones.

There are 14 phalange bones in total to each foot – the same as in each hand.

There are plane joints between the individual tarsal bones as well as between the tarsals and the metatarsals; these allow slight gliding movement only. Between metatarsals and phalanges there are condyloid joints and between individual phalange bones there are hinge joints.

The arches of the feet

The feet have three major arches that distribute the weight of the body between the ball and the heel when the body is standing or walking.

1. **Medial longitudinal arch** – running along the inside of the foot

2. **Lateral longitudinal arch** – running along the outside of the foot

3. **Transverse arch** – running across the foot.

The bones of the feet, the ligaments connecting them and the muscles of the feet maintain the shape of these arches.

Functions of the skeletal system

Now that you have tested the structures of your own skeletal system, it will be useful to look in more detail at exactly what the system can do as a whole.

The skeletal system has five main functions including protection, support and shape, movement, storage and the production of blood cells.

Protection

Bone protects the internal organs e.g.

- The skull protects the brain.
- The spine protects the spinal cord.
- The thorax protects the heart and lungs.
- The pelvic girdle protects the reproductive organs.

Support and shape

The bones give the body its unique shape and the skeletal system as a whole is well equipped to be weight bearing e.g.

- Bone supports the weight of the whole body i.e. skin, muscles, internal organs and excess fatty tissue!
- Cartilage offers shape to structures such as the ear and nose as well as providing additional support between the bones at joints e.g. spine.
- Ligaments hold bones together at joints adding increased support.

Movement

The skeleton acts as a framework for the attachment of muscles i.e.

- Tendons attach muscles to bone.
- Contraction of muscles moves the bones within a range of movement determined by the type of joint e.g. maximum movement at the ball and socket synovial joint of the hip.

Storage

The blood takes organic matter from the blood i.e. minerals and fat to be stored in the cavities within bones e.g.

- Calcium and phosphorus – if the level of these minerals in the blood is greater than is needed, the excess is stored in the bones contributing to their strength and resilience. If the levels within the blood decrease, the bones release their store back into the blood.
- Fat – this is also stored in the bones as yellow bone marrow and is released into the blood stream when needed and used by other parts of the body as a source of energy.

Production of blood cells

The red bone marrow found in the cancellous bone tissue is responsible for producing new blood cells.

The study of the skeletal system provides us with an insight into the way in which all parts of the body work together as a whole. Always remember to link systems with the other systems with which they interact, never forgetting that they cannot work in isolation!

Common conditions

An A–Z of common conditions affecting the skeletal system

- ANKYLOSING SPONDYLITIS – a disease of the joints usually affecting the spine and resulting in back pain and stiffness.
- ARTHRITIS – inflammation of the joints. Arthritis may be acute or chronic.
- BUNION – a swelling of the joint of the big toe made worse by excessive pressure.
- BURSITIS – the bursa becomes inflamed and affects the movement within the joint. When this disorder affects the knee it is known as *housemaid's knee*.
- CARTILAGE, torn – a knee injury resulting from sudden twisting movements which tear the cartilage that lies between the joints.
- CHONDROSARCOMA – tumours that grow slowly and are generally the result of malignant change in benign tumours.
- COCCYDYNIA – pain in the base of the spine usually occurring after injury to the coccyx bone.
- DUPUYTREN'S CONTRACTURE – a fixed bending of the fingers due to the shortening and thickening of fibrous tissue in the palm of the hand.
- FLAT FEET – lack of arching in the feet causing foot strain and pain.
- FRACTURE – a break or a crack in a bone due to injury, repeated stress to a bone or weakened bone caused by disease.
- FROZEN SHOULDER – a severe aching pain in the shoulder affecting the middle-aged and elderly, which restricts shoulder movements.
- GANGLION – a harmless swelling at a tendon near a joint. Usually found on the hands or feet.
- GOUT – pain in the joints, particularly the big toe, are a symptom of this condition which is an upset of the chemical processes of the body. Knees, ankles, wrists and elbows may also be affected.
- KYPHOSIS – concave curve of the spine in the thoracic region creating a 'hump back'.
- LORDOSIS – convex curve of the spine in the lumbar region creating a 'hollow back'.

- MALLET FINGER – a finger that cannot be straightened due to damage to the tendon.
- METATARSALGIA – pain along the ball of the foot common in middle-aged and overweight people.
- OSTEOARTHRITIS – a degenerative disease of the joints. The cartilage within the joints wears away resulting in pain. Extreme cases result in a joint replacement e.g. hip, knee etc.
- OSTEOCHONDRITIS – softening of bone causing the bone to change shape and become deformed. Affects children.
- OSTEOGENESIS – defect of the bone cells causing 'brittle bones'.
- OSTEOMALACIA or rickets – softening of bone due to lack of Vitamin D.
- OSTEOMYELITIS – Inflammation of the bone caused by a bacterial infection often as a result of a localised injury.
- OSTEOPOROSIS – weakening of the bones which may be caused by changing levels of the hormones oestrogen and progesterone.
- OSTEOSARCOMA – a rapidly growing malignant tumour of the bones.
- PAGET'S DISEASE OF THE BONES – thickening of the bones causing pain and broadening of bone.
- RHEUMATOID ARTHRITIS – a destructive swelling of the joints initially affecting the fingers and feet, then spreading to the wrists, knees, shoulders, ankles and elbows
- SCOLIOSIS – a lateral (away from the mid line) curve in the spine.
- SLIPPED DISC – a bulging of one of the fibrocartilage discs that separate the vertebrae causing pain and muscle weakness.
- SPRAIN – a sudden stretching or tearing of a ligament resulting in pain and swelling.
- STRESS – stiff joints and repetitive strain injury are symptoms of the effect of excessive stress on the skeletal system.
- SYNOVITIS – inflammation of a joint after injury.
- WHIPLASH – backward jerking of the neck resulting in damage to the spine.

System sorter

SKELETAL SYSTEM

Muscular

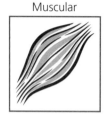

Respiratory

Integumentary

The vitamin D produced in the skin helps the bones absorb calcium which is needed to keep them strong and healthy. A lack of vitamin D in children results in the malformation of bones – a condition known as rickets.

Tendons attach muscle to bones. Muscles pull on bones at synovial joints causing movement. The range of movement is dependent on the type of joint ie ball and socket joint of the shoulder and hip allow maximum movement.

The axial skeleton comprises of the bones of the skull, spine and thorax. It is the bones of the thorax that form the rib cage offering protection to the vital organs of the respiratory system.

Circulatory

Red bone marrow is found in cancellous bone tissue and is responsible for the formation of new blood cells. The appearance of red bone marrow is greater in childhood reducing with age.

Calcium and phosphorus present in the food we eat as part of a healthy diet is processed by the digestive system before being transported to the bones of the skeleton by the blood. Calcium and phosphorus are responsible for strong healthy bones.

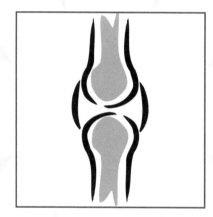

Hormones released by the endocrine glands into the blood stream are responsible for controlling the growth rate of bones in children. Hormone imbalance is also responsible for the changes in bone density experienced in women after the menopause.

Any excess of calcium and phosphorus in the body that is not stored in the bones is excreted from the body in urine.

The bones of the skull and spine that make up the axial skeleton protect the delicate nervous tissue of the brain and spinal cord.

Digestive

Genito-Urinary

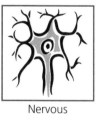

Nervous

Endocrine

The skeletal system consists of active living tissue that helps to give the body its shape and form. The bones of the axial and appendicular skeleton protect vital organs and work with other body systems to create movement. This system helps to control calcium and phosphorus levels in the body and is responsible for the production of new blood cells in the red bone marrow. In turn, this system relies on its links with the other systems of the body for its own survival.

Holistic harmony

The skeletal system is a complex network of organs that contribute to the external and internal well-being of the body as a whole. The skeleton provides us with a basic shape to which muscles (Chapter 4) and skin (Chapter 2) are added, making up an external framework that distinguishes us as being part of the human race, but also sets us apart from other people because of our own individual external characteristics. Internally, the skeletal system maintains vital links with the other systems to ensure that its functions of movement, protection, production and storage take place efficiently and effectively. It is very easy to take these functions for granted and an appreciation of how the body gets it right and what happens when it gets it wrong often prompts us to take more responsibility for our own well-being. There are many things we can do to help the skeletal system function better for longer and these include maintaining the fundamental balance between internal and external care.

Fluid

Approximately 25 per cent of bone is water and the synovial fluid bathing the joints is also comprised of water. Most of this water comes from the fluid that we drink and the fluid in the food that we eat e.g. fruit and vegetables. This fluid goes from the digestive system into the blood and is then taken to the bones. It is therefore important to ensure that we maintain the fluid levels within the body by drinking adequate amounts of fluid. We need to appreciate the fundamental differences between 'good' and 'bad' fluid. Simple water is what constitutes 'good' fluid, the benefits of which must never be underestimated for it is vital to life. Fluid is not so good and even 'bad' when we add other things to it, especially caffeine. Caffeine is found in tea, coffee and cola drinks and acts as a **diuretic** meaning that it increases the production of urine thus lessening the rehydration of the body through drinking. Lack of water in the skeletal system will contribute to dry, brittle bones and stiff joints that are more vulnerable to damage as a result.

Nutrition

Bones are constantly being rebuilt as old bone is broken down by osteoclasts and new bone is formed by osteoblasts, and so therefore they are extremely sensitive to nutritional

Remember

Remember that Vitamin D is also produced in the skin in response to sunlight.

Angel advice

It is important to remember that the body as a whole needs a balance of all foodstuffs but that certain foods have more of an effect on certain systems.

factors. Therefore, for the skeletal system to maintain its functions, it relies on a good, healthy diet containing:

- Calcium – found in Swiss and cheddar cheeses and improves bone density.
- Magnesium – found in almonds and cashew nuts and helps to strengthen bones.
- Phosphorus – found in most foods and helps in the formation and development of bones.
- Vitamin D – found in fish such as herrings, mackerel and salmon, and helps bones to absorb and store calcium.
- Vitamin C – found in peppers, watercress and cabbage and is responsible for collagen production, which keeps bones and joints firm and strong.
- Zinc – found in pecan nuts, Brazil nuts and peanuts and helps to make new bone cells.

Research has shown that an excessively high protein diet can lead to calcium deficiency because proteins are acid forming and calcium is neutralising. The higher the intake of proteins, the greater the need for calcium which is taken away from the bones leading to weakened bones in the long term. This is thought to be a primary cause of osteoporosis.

The battle against **free radicals** continues in the skeletal system and antioxidant nutrients in the form of vitamins A, C and E help to combat this activity and prevent damage to bone tissue.

Rest

It is important to get the right balance between rest and activity in order to maintain a healthy skeletal system. Getting the balance wrong can lead to:

- Stiff joints resulting in restricted movement.
- Wasted or weakened bones resulting in loss of strength.

Activity

The skeletal system will naturally develop greater strength in weight-bearing bones, but at the same time will lose bone mass in bones that are not used i.e.

- Athletes are able to build up the bones most needed to perform their sport by maintaining a high mineral content in the individual bones making them stronger.

- People who are bedridden find that their bones weaken and waste due to a loss of minerals. The same occurs if a bone is set in a cast. The bones will need building up and this can be done through exercise.

The body is able to detect its own needs and respond accordingly by holding on to or releasing calcium. There is however a limit to this because:

- excessive exercise can result in damage to bones and joints if there is inadequate resting time just in the same way as
- excessive resting will result in eventual lack of movement if there is inadequate activity by the body!

Air

Sensitivities can have an effect on the skeletal system, for example many people experience sensitivity to gas and exhaust fumes. When breathed into the body, these factors reduce the effectiveness of the skeletal system resulting in a higher risk of conditions such as rheumatoid arthritis and osteoarthritis and a worsening of the symptoms for those people already suffering from these conditions. Care should be taken to avoid excessive exposure to pollution e.g. car exhaust fumes, cigarette smoke etc. Breathing in fresh, clean air ensures that the body receives the right amount of oxygen needed to 'feed' the bones and activate the energy required to perform the chemical reactions in order for the skeletal system to function.

Age

The ageing process is responsible for the slowing down and eventual destruction and death of the cells that make up the human body. We cannot live forever and we cannot maintain a young body indefinitely because of the various functions of the body that are beyond our control. The skeletal system contributes to this process with a gradual reduction in its efficiency to perform vital functions, as bones decrease in strength and joints lose their flexibility. We therefore have a limited time span in which to use our bodies although it is gradually increasing as we become more aware of our body's needs. Human life expectancy has increased as our world has

Angel advice

- **T**ake the time to learn what it is your body needs!

- **I**ndulge in time to look after your body to meet those needs!

- **M**otivate your mind to take action on your body!

- **E**njoy it all while you can!

developed and there are more opportunities open to us to help ourselves.

Colour

The axial skeleton is the site for the seven primary **chakras**. The term 'chakra' originates from Indian tradition and is a Sanskrit word meaning 'wheel'. Chakras are classified as wheels of light that attract energy. This energy is thought to radiate externally and internally merging with the body systems and producing an effect on our general well-being. Each chakra is associated with different parts of the body and is identified by a colour. At a basic level, the relative anatomical position of the chakra provides a guide to the body parts they assist, and the associated colour reflects the order of those found in a rainbow:

- The first chakra is located in the coccyx region of the spine and its associated colour is red.
- The second chakra is found in the sacral region of the spine and is associated with the colour orange.
- The third chakra is located in the lumbar/thoracic region of the spine and its associated colour is yellow.
- The fourth chakra is found in the higher thoracic region of the spine in line with the sternum. The associated colour is green.
- The fifth chakra is located in the cervical region of the spine and is associated with the colour blue.
- The sixth chakra is located in the centre of the forehead and is associated with the colour indigo.
- The seventh chakra is found in the middle of the head and the associated colour is violet.

When the human body is well and happy, these wheels are able to rotate freely with energy helping to maintain holistic harmony. However, stress and illness are believed to cause blockages of energy within the chakras which can be counteracted by the use of a corresponding or complementary colour. For example public speaking is a very daunting task and one which affects the throat region; the throat is in the area of the fifth chakra and its associated colour is blue, so wearing a blue scarf or a blue necklace around the neck can help to energise this area, which in turn aids the task of speaking in public. To the uninitiated this may seem a bit odd, but it is worth exploring as it may prove to be a safer and cheaper alternative to more conventional methods of stress relief.

Awareness

Studies have shown that our state of mind contributes greatly to the state of our bodies, proving that a 'happy' mind results in a 'happy' body.

Being happy relies on being accepted, not just by others but perhaps more importantly being accepted by ourselves! How many times do we hear ourselves saying 'I dislike my size, my shape, my height'? These are all factors that are controlled by the skeletal system and we can develop a very negative view of its functions by hating those things about us. We cannot change our bone structure in any radical way so it is up to us to live with our own unique skeletal system and learn to accept it. After all it provides us with so much in terms of movement and protection!

Negative thoughts develop negative feelings, which in turn develop physical negatives in the form of disorders and disease. Anger, fear and hatred can all manifest themselves in physical symptoms that have a harmful influence on the body's well-being. Be aware of the contribution your skeletal system has made to your turning over the pages of this book, to your position in the chair you are sitting in and to the tasks it allows you to do as part of your working day. Amazing isn't it?

Special care

The skeletal system's response to excessive stress can have a serious effect on its well-being and this highlights the need to ensure that a balance is maintained between external and internal factors in order to keep the skeletal system in optimum health.

External stress:

- Excessive exercising resulting in strain and damage.
- Excessive repetitive movements resulting in **RSI**, repetitive strain injury.

Internal stress affects the balance of hormones responsible for maintaining healthy bones:

- Childhood – time of maximum development of bones when hormone balance affects growth rate.

- Puberty – time of enormous change when hormone balance affects the development of the skeletal system into adulthood.

- Pregnancy – hormone release is needed for the development of the baby and the maintenance of the mother.

- Menopause – changes in hormone levels resulting in a lack of hormones have a direct weakening action on the skeletal system.

- Excessive emotional strain – hormones released to initiate a 'coping mechanism' can have a long-term detrimental effect on the skeletal system e.g. the digestive system may be suffering as a result of a lack of nutrients being taken to the bone cells which will in turn affect the efficient renewal of bone tissue.

Meeting the needs of our skeletal system is therefore vital if we are to keep the body functioning properly and the special care associated with the practice of **safe stress** is a good start!

Remember

Beauty and holistic treatments provide the ideal basis for safe stress and should be regular and progressive. Use the Treatment Tracker for guidance.

Treatment tracker

SKELETAL SYSTEM

Make up

Corrective make up techniques can be used to enhance bone structure. Highlighting above and shading below the zygomatic bones will create the illusion of high cheek bones.

Facials

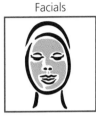

Petrissage movements used during facial massage can help to drain the sinuses of the facial bones freeing the mucus that sometimes causes problems.

Nail care

Remedial hand and feet treatments such as paraffin wax aids joint flexibility and mobility by warming and soothing the synovial fluid that bathes the joints.

Hair removal

The shadows created by the excessive dark hair of the eyebrows can cause the bone structure of the eyes to appear heavy. Removal of this hair creates a lighter effect which is more flattering.

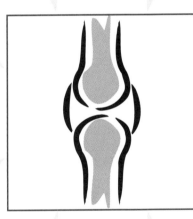

Friction movements warm the synovial fluid bathing the joints. Joint manipulation movements may then be performed to improve mobility and flexibility.

Audio sonic vibration used around the synovial joints of the body helps the soreness associated with repetitive strain injury by stimulating blood flow to the area.

Working over the reflex points relating to the spine and the joints will aid mobility and help to ease away the stress associated with stiff joints.

Painful joints may be treated with a blend of camomile, juniper and rosemary mixed with a carrier oil.

Electrical

Massage

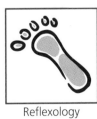
Reflexology

Aromatherapy

The skeletal system gains direct benefit from beauty and holistic treatments aiding cellular function and joint mobility. Bones remain strong and healthy and joints stay more flexible.

Knowledge review – Skeletal system

1   What are bone-forming cells called?

2   What do the cells osteoclasts do?

3   What do the axial and appendicular skeleton consist of?

4   What type of bones make up the arms and the legs?

5   What is the correct term that describes freely movable joints?

6   What type of joint is formed between the bones of the cranium?

7   What is the main inorganic substance stored in the bones?

8   What type of bone makes up the length of long bones?

9   Where are red and yellow bone marrow found?

10 What is the function of red bone marrow?

11 What is yellow bone marrow?

12 Why are vitamins C and D vital for the development of bone?

13 How do hormones affect the bones?

14 What are the functions of ligaments and tendons?

15 What type of joint allows a maximum range of movement?

16 Which type of joint allows only a gliding movement between bones?

17 What is it that covers bones along their length?

18 What covers bones at a joint to prevent friction?

19 Name the bones that make up the cheek, temple and jaw on the face.

20 Name the areas of the spine and state how many bones there are in each area.

The muscular system

4

Learning objectives

After reading this chapter you should be able to:

- **Recognise the different types of muscular tissue**

- **Identify the main skeletal muscles**

- **Understand the functions of the muscular system**

- **Be aware of the factors that affect the well-being of the muscular system**

- **Appreciate the ways in which the muscular system works with the other systems of the body.**

During the next phase of our journey through the human body, we are going to look closely at the way in which the body moves. We can already identify the bones, which contribute to moving the body, and we can appreciate the range of movements made possible by the synovial joints (Chapter 3). Now we are going to learn how the actual movement takes place and the system responsible for this is **THE MUSCULAR SYSTEM.** This system consists of different types of muscular tissue all responsible for some form of movement. Muscles function like motors by producing the force to make a movement of the external and internal parts of the body. Like motors, muscles need fuel and they take this from the nutrients brought to them via the blood from the digestive system (Chapter 7) and respiratory system (Chapter 5) whilst the nervous system (Chapter 9) co-ordinates these muscular activities.

Science scene

An example of this voluntary and involuntary action would be the swallowing of food, which is voluntary, and the passage of food through the digestive system, which is involuntary.

Structure and position of muscles

There are many thousands of muscles making up the human body. Some are arranged internally whilst others are attached to bones, skin or other muscles in order to ensure that specific movements can be executed effectively. Muscles move either voluntarily at will – that is we can control the movement – or involuntarily without any conscious control depending on their tissue type.

There are three types of muscular tissue. These include **cardiac**, **visceral** and **skeletal** muscular tissue:

Cardiac muscular tissue

Cardiac muscular tissue is exclusive to the heart. It is striated in appearance with each cell containing a nucleus. It causes the heart to expand and contract, pumping blood through the blood vessels and around the body. It is under involuntary control.

Visceral muscular tissue

Visceral muscular tissue is also known as **smooth** tissue because of its appearance. The cells are spindle-shaped and form into bundles. Each cell contains a nucleus but has no distinct membrane so appears smooth. It is under involuntary control and is found internally. Visceral muscle tissue is responsible for moving food through the digestive system (Chapter 7) and waste through the urinary system (Chapter 8). The arrector pili muscles in the skin (Chapter 2) are also visceral muscles which contract when the body temperature changes producing 'goose bumps'. These actions occur without any conscious effort on our part.

Skeletal muscular tissue

Skeletal muscular tissue is also known as **striated** tissue because of its stripy appearance. It is responsible for taking the human body through a range of movements which are under voluntary control. It is the skeletal muscles that need to be examined further by the therapist in order to perform effective treatments. Skeletal muscles are arranged both deeply and superficially within the body depending on their function and action. Many muscles overlap one another and many muscles work together to make a single movement.

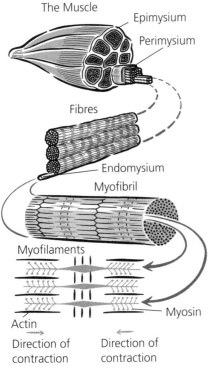

The Muscle
Epimysium
Perimysium
Fibres
Endomysium
Myofibril
Myofilaments

Actin
Myosin
Direction of contraction   Direction of contraction

Skeletal muscle structure

Fascinating Fact

Muscles can make up approximately 50 per cent of the total body weight depending on sex, age and fitness levels.

Structure of skeletal muscles

Muscles are made up of active living tissue containing:

- Water – approximately 75 per cent
- Inorganic substances – approximately 5 per cent e.g. mineral salts
- Organic substances – approximately 20 per cent including muscle forming cells called **myoblasts** together with a blood and nerve supply.

Muscle formation

Myoblasts are responsible for forming muscle fibres, which collectively form the muscles themselves. The number of muscle fibres present in a body is relatively constant after birth as they have the ability to grow and increase in size. Muscle fibres contain threadlike structures called **myofibrils**, which extend from one end of the fibre to the other end. Each myofibril is made up of even smaller threads or *filaments* of protein called **myofilaments**. There are two types of myofilaments:

1. **Actin** – thin filaments
2. **Myosin** – thick filaments.

Also present in the muscle fibres are **mitochondria**. These are often referred to as 'power houses' because they are responsible for generating the energy needed for the muscle to make a movement, which in turn moves the body. These 'power houses' store **glycogen** and **myoglobin**. Glycogen is the end product of the carbohydrates that we eat and is needed to create energy, whilst myoglobin holds the oxygen brought to the muscles from the respiratory system (Chapter 5) as we breathe in, and is needed to activate the energy.

The muscle fibres are surrounded by connective tissue called **endomysium** which provides support. Groups of muscle fibres make up bundles which are surrounded by more connective tissue called **perimysium** which provides further support. Bundles of muscle fibres are arranged in groups to form a complete muscle, which is covered by a further layer or **fascia** of connective tissue called **epimysium**. It is because of this structure that skeletal muscles have a striated or striped effect – like a bundle of elastic bands.

Muscles contain a rich blood supply to ensure that the fibres are provided with 'fuel' and a nerve supply linking the muscular system to the brain in order to allow instructions for movement to be carried out.

1 spindle - shaped

2 flat

3 triangular

4 circular

Muscle shapes

!

Remember

These sphincture muscles are not fully operational in babies hence the need for nappies! Loss of control in these muscles in later life results in incontinence.

Muscle shape

Skeletal muscles come in four different basic shapes:

1. Spindle-shaped – thick middle section with tapered ends e.g. biceps and triceps muscles in the upper arm

2. Flat – thin sheets of muscle e.g. frontalis muscle of the forehead

3. Triangular – wider at one end coming to a point at the other end e.g. deltoid muscle of the shoulder

4. Ring-shaped – surrounding an opening e.g. orbicularis muscles of the eyes and mouth, and **sphincter** muscles which surround the external openings of the anus and the bladder, and close off the final sections of the digestive and renal systems at will.

Muscle attachment

Most muscles are attached to **bones** by tough, fibrous connective tissue called **tendons** (Chapter 3) at an **origin** at one end and an **insertion** at the other end.

● The origin is where the muscles are attached at a *fixed* point.

● The insertion is generally where the muscles are attached at a joint making it a *movable* point.

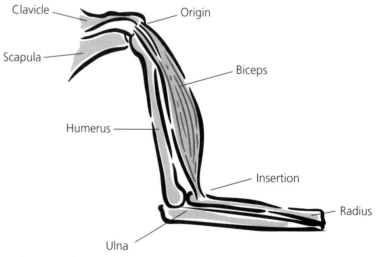

Attachment of muscles

Many of the small facial muscles attach to one another or to the facial skin itself pulling the face a specific way as they contract, forming individual facial expressions.

Muscle contraction

There are two types of muscle contraction:

1. **Concentric** contraction – shortening of the muscle
2. **Eccentric** contraction – lengthening of the muscle.

When a muscle performs a concentric contraction, the actin and myosin filaments overlap pulling the muscle fibres together from their length and making a bulky middle, in much the same way as an extending ladder shortens. The opposite occurs when a muscle contracts eccentrically, extending the filaments away from each other.

Because the muscular tissue is elastic, it is able to return to its original length following a contraction. The force of the contraction is dependent on the amount of fibres contracting at the same time. The greater the numbers of contracting muscles, the greater the force.

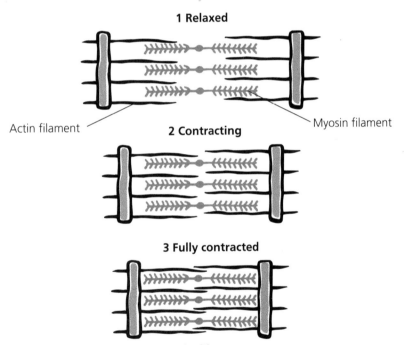

Resting and contracting muscle fibres

Fascinating Fact

When muscles contract very forcefully, they can generate up to 50 pounds of pull for each square inch of muscle! If the force is too great the muscle tendons may break away from the bones.

Muscles are made up of different types of fibres allowing varied types of movement:

- **Slow twitch** fibres – produce slow contractions with low levels of speed and power over long periods of time
- **Fast twitch** fibres – provide rapid contractions with high levels of speed and power for only a short period of time.

Long distance runners are born with more slow twitch fibres making them more suited to running marathons. Sprinters on the other hand have a much higher proportion of fast twitch fibres enabling them to be more effective at running short distances in a very short amount of time.

The human body contains varying amounts of both muscle fibres depending on genetic factors and this cannot be changed, although the individual action of each type of fibre can be improved through exercise.

Types of muscles

Depending on their action, muscles are either **agonists** or **antagonists** and either **synergists** or **fixators.**

- Agonists or **prime movers** are muscles where the action (contraction) takes place.
- Antagonists are the muscles which lie opposite to the agonist muscles and work in an *opposing* manner i.e. they relax as the agonists contract.
- Synergists are small muscles which *help* the agonist to ensure that a more controlled action can take place.
- Fixators are the large muscles responsible for maintaining posture. Their action is to *fix* the position of the body when a specific movement takes place.

Blood supply to the muscles

Muscles need nutrients and oxygen which are brought to them via the blood to 'fuel' the action of movement. This is carried out in three ways:

Remember

The oxygen debt is the amount of oxygen needed to restore ATP in the muscle fibres and the amount of oxygen needed by the liver cells to convert the lactic acid to glycogen.

1. Glycogen stored in the mitochondria in the muscles from the digestive system, plus oxygen in the myoglobin brought to the muscles from the respiratory system, create **ATP** (**A**denosine **Trip**hosphate) by oxidation. ATP is the chemical fuel needed to make a movement in the muscle. As a result of this action, waste product called **pyruvic acid** is produced, which is utilised by the oxygen to create more energy. This is called the **aerobic** energy system. Other waste products including carbon dioxide and water are produced which are carried away by the circulatory system to be released out of the body.

2. When the need for energy is high, more pyruvic acid is produced than the oxygen can cope with and the excess turns into lactic acid. The lack of oxygen creates an oxygen debt in the muscles and lactic acid builds up causing fatigue and resulting in breathlessness and aching muscles. This is known as the **anaerobic** energy system. When the need for energy reduces, the body is

Fascinating Fact

Some hours after death, the muscles contract and fix the joints. This is called **rigor mortis** and is due to a chemical reaction which prevents the relaxation of muscle fibres. The fibres remain contracted until the muscles start to decompose.

Remember

The motor nerves coming from the brain tell the muscles to move when we want them to (like starting up a motor) and the sensory nerves tell the brain to make the movement stop when we feel the muscles start to ache (switching off the motor).

able to take deep breaths, thereby bringing more oxygen to the muscles and repaying the oxygen debt. Carbon dioxide, water and lactic acid are removed from the muscles by the circulatory system. The lactic acid is taken to the liver and broken down into glycogen to be used as energy at a later date. This combined process gradually allows aching muscles to return to normal.

3. There is a small amount of energy already stored in the mitochondria in the muscles in the form of ATP which can be used for short bursts of energy only before one of the other methods 'kicks in'.

Nerve supply to the muscles

For muscles to contract, they are stimulated through the nervous system by **motor** and **sensory** nerves:

1. Motor nerves from the brain enter the bulky centre of the muscle at a **motor point**. The nerves then branch out with **motor end plates** attaching to each muscle fibre. The motor point receives messages from the brain which are then relayed to each fibre informing the muscle as a whole to contract or relax.

2. Sensory nerves run parallel to the motor nerves back to the brain reporting on the action within the muscle.

Muscle development

The amount of muscle fibres present in the body remains constant throughout life but we are able to increase muscular strength, flexibility and endurance through exercise and reduce the level of performance when a muscle is not used. Muscle fibres will increase in size through regular use and decrease as a result of lack of use.

If muscle fibres are damaged accidentally or cut for example during surgery, the damaged tissue is removed by **phagocytosis**. Cells called phagocytes engulf the damaged tissue, which is then replaced.

Where there is minor damage to a muscle, the surviving muscle fibres develop outgrowths to make up for the lost tissue and the muscle is completely restored. Where the damage is greater, the fibres are unable to make up for the lost tissue and fibrous or scar tissue forms inhibiting the action within the muscle which may in turn restrict the joint movement.

Posterior view   Anterior view

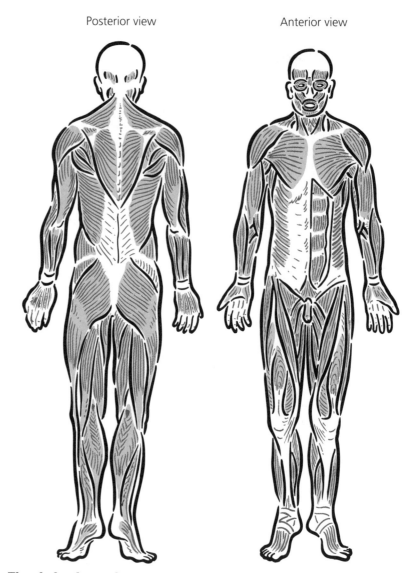

The skeletal muscles

The muscles of the body

A therapist needs to be able to identify the different types of muscular tissue that make up the human body and in particular those muscles with which they have direct contact. This knowledge will enable a therapist to adapt the application of treatment to suit the needs of the individual client. There are therefore great benefits to be had by having an awareness of certain individual muscles and groups of muscles. These include:

Muscles of the facial expression:

* **Occipito-frontalis** – covering the occipital bone at the base of the cranium and the frontal bone at the front of the cranium, it forms the forehead, raises

---

Fascinating Fact

There are over six hundred named muscles in the body! Fortunately, as therapists we do not need to know them all! It is important, however, to have a working knowledge of the main skeletal muscles as described in this chapter and appreciate that many more muscles contribute to the intricate set of movements that the human body is capable of making.

the eyebrows in surprise and is responsible for horizontal frown lines.

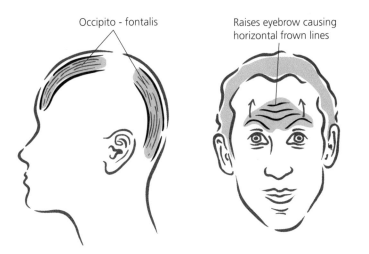

Occipito - fontalis

Raises eyebrow causing horizontal frown lines

- **Corrugator supercilli** – situated between the eyebrows drawing them together creating a vertical frown line.

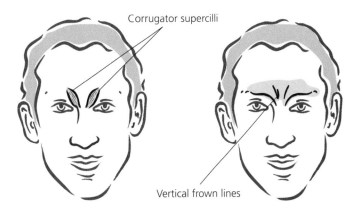

Corrugator supercilli

Vertical frown lines

- **Orbicularis occuli** – a circular muscle surrounding the eye. It is responsible for closing the eye and contributes to the fine lines that first appear when you close your eyes tightly, developing into crow's feet over time.

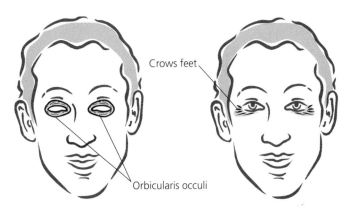

Crows feet

Orbicularis occuli

- **Zygomaticus** – covering the zygomatic bones and attached to the muscles of the mouth. Responsible for lifting the mouth and the cheeks as we laugh.

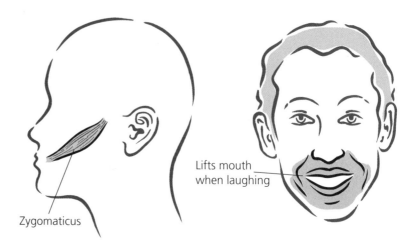

Lifts mouth when laughing

Zygomaticus

- **Risorius** – situated in the lower cheek area and is attached to the corners of the mouth lifting the mouth in a smile extending to a grin.

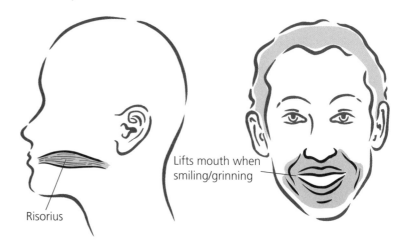

Lifts mouth when smiling/grinning

Risorius

- **Buccinator** – situated in the cheek area between the maxilla and mandible bones and is responsible for squeezing the cheeks together as in blowing and chewing.

Puffs out cheeks
when blowing/
chewing

Buccinator

● **Nasalis** – covers the front of the nose and causes wrinkling of the nose when the muscle is compressed.

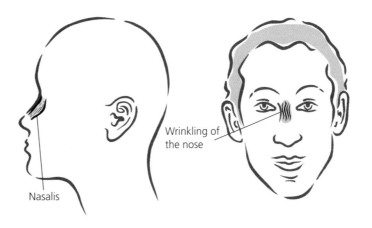

Wrinkling of
the nose

Nasalis

● **Procerus** – covers the bridge of nose drawing the eyebrows downwards puckering up the skin of the top of the nose into transverse wrinkles.

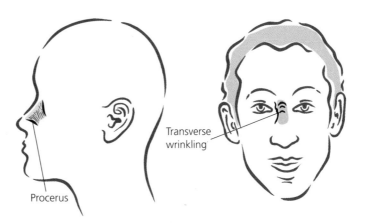

Transverse
wrinkling

Procerus

- **Orbicularis oris** – a circular muscle surrounding the mouth forming its shape during use. Responsible for puckering and compressing the mouth as in kissing.

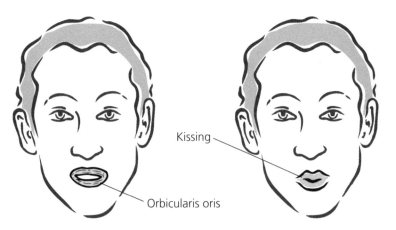

Kissing

Orbicularis oris

- **Triangularis** – extends along the side of chin drawing the angles of the mouth down as in sulking.

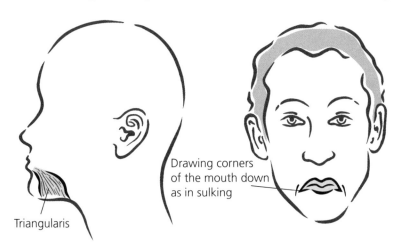

Drawing corners of the mouth down as in sulking

Triangularis

- **Mentalis** – situated at the top of chin and is responsible for raising the lower lip as in doubt or displeasure causing the chin to wrinkle.

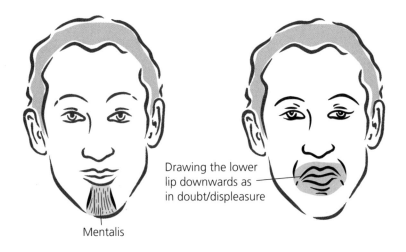

Drawing the lower lip downwards as in doubt/displeasure

Mentalis

Fascinating Fact

Mastication refers to the action of chewing.

Fascinating Fact

Grinding the teeth, a common stress related disorder is responsible for 'jaw ache'!

Muscles of mastication:

- **Temporalis** – situated at the side of the head from the ear to the mandible. Responsible for lifting the mandible and drawing it back when chewing.

- **Masseter** – situated in the cheek from the zygomatic to the mandible bones. It raises the mandible and is responsible for closing the mouth and clenching the teeth.

- **Buccinator** – situated between the upper and lower jaw drawing the cheeks together as we chew.

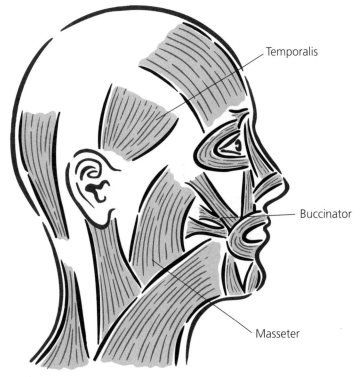

Muscles of mastication

Muscles of the neck, back and chest:

- **Platysma** – a broad muscle covering the front of neck from the chin to the chest. It draws the jaw and lower lip downwards as in sadness and contributes to the necklace lines on the neck.

- **Sterno cleido mastoid** – extends from the temporal bone down to the sternum and clavicle bones at the sides of neck. The two muscles work together to bring the head forward and in opposition to move the head from side to side.

- **Trapezius** – a large triangular muscle covering the back of the neck and upper back. Works with the sterno cleido mastoid to move the head to either side

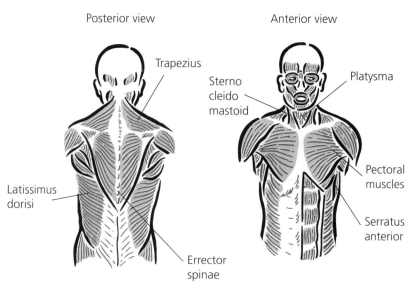

Muscles of the neck, back and chest.

and against the platysma to draw the head back. Together the trapezius and platysma muscles cause the head to nod. It also contributes to shoulder movements.

- **Errector spinae** – group of muscles, which extend along the spine at the centre of the back from the neck to the pelvis. Responsible for maintaining upright posture and extending the spine.

- **Latissimus dorsi** – covers either side of the back from the underarms to the lumbar region and is used in rowing and climbing, working the shoulder joint.

- **Pectoralis major** and **minor** – covering the chest area under the breasts. The pectoralis minor lies directly below the major and together they contribute to the shoulder movements involved with throwing and climbing.

- **Serratus anterior** – situated under the arms contributing to shoulder movements used when pushing and punching.

Muscles of respiration (breathing):

- **Diaphragm** – large dome-shaped muscle separating the thorax from the abdomen. Flattens on contraction to increase the area in order to allow the lungs to fill with air as we breathe in. Returns to normal shape as we breathe out.

- **Intercostals** – internal and external, situated in between the ribs forming the shape of the thorax. These muscles work together to increase the space in

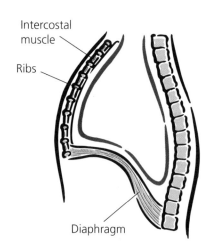

Muscles of respiration

Activity

Remember, muscles do not work in isolation to create movement. Think about the different muscles needed to contribute to a specific action such as blowing or eating. Try the action and work out which muscles enabled you to carry out the movements. Now think about the contribution from the other body systems required to help you to make those movements. This analysis helps us to appreciate the complexity of the human body and to understand the intricate links it makes within itself!

! Remember

The biceps is the agonist muscle or prime mover, the brachialis the synergist, the helper, and the triceps the antagonistic muscle working against the agonist.

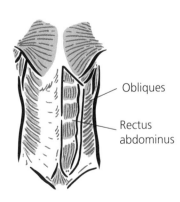

Muscles of the abdomen

the thorax (external intercostals) when breathing in and to depress the ribs (internal intercostals) when breathing out, coughing and blowing.

Muscles of the shoulders and arms:

- **Deltoid** – covers the top of the arm and shoulder from the clavicle to the upper part of the humerus. Assists in the movement at the shoulder joint, lifting the arms up, back and forwards.

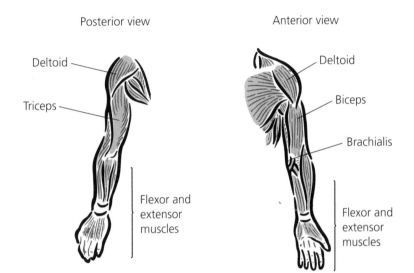

Muscles of the shoulders and arms

- **Biceps** – situated at the front of the upper arm and is responsible for flexion of the arm at the elbow and supination of the forearm and hand turning the palm upwards.
- **Triceps** – situated at the back of the upper arm and works in opposition with the biceps to extend the arm.
- **Brachialis** – situated at the front of arm below biceps and works with the biceps to flex the arm.
- **Flexor** and **extensor** muscles – situated in the forearm, hand and fingers and responsible for bending and straightening the wrist, hand and finger joints.

Muscles of the abdomen:

- **Rectus abdominus** – situated in the centre abdomen from the sternum to the pelvis. Works in opposition with the erector spinae muscles to flex the spine and compress the abdomen and contributes to maintaining upright posture.

Tip

Exercising the rectus abdominus by doing sit ups contributes to the 'six pack' body shape and the curve of the waist can be enhanced by exercising the oblique muscles with waist twists and side bends.

- **Obliques** – external and internal muscles forming the waist. This set of muscles lie obliquely on either side of the rectus abdominus with the external obliques facing inwards and the internal obliques facing outwards. They allow the trunk to move from side to side.

Muscles of the hips and legs:

- **Gluteal** muscle group – gluteus maximus, medius and minimus forming the hips and buttocks from the pelvis to the femur. Responsible for movement in the hip joint – walking, running and maintaining an upright position. Also known as the abductor group of muscles as they move the leg away from the centre line of the body.

- **Adductor** muscle group – four muscles of the inner thigh. Contribute to movements at the hip joint and move the leg towards the centre line of the body.

- **Hamstring** muscle group – a group of three muscles forming the back of the thigh from the pelvis to below the knee. Responsible for flexing the knee and extending the hip backwards as needed for walking and jumping.

- **Quadricep** muscle group – a group of four muscles forming the front of the thigh lying opposite to the hamstrings. They work in opposition to the

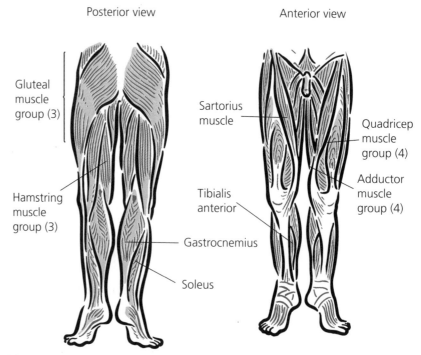

Posterior view   Anterior view

Gluteal muscle group (3)

Hamstring muscle group (3)

Sartorius muscle

Quadricep muscle group (4)

Adductor muscle group (4)

Tibialis anterior

Gastrocnemius

Soleus

Muscles of the hips and legs

Activity

Work out which of these three muscles acts as the agonist, the synergist and the antagonist when you stand on tiptoes: gastrocnemius, tibialis anterior and soleus.

You will find the answer on the next page.

hamstring muscles extending the knee and flexing the hip as needed for walking and kicking.

- **Sartorius** – crosses the front of the thigh from the outer side of the pelvis to the inner side of the knee. Assists with movements at the hip joint and is used to turn the leg outwards.
- **Gastrocnemius** – forms the bulk of the calf at the back of the lower leg. Assists in walking and running by flexing the knee and plantar flexing the foot – bending the foot downwards, providing the 'push' needed for such movements.
- **Tibialis anterior** – forms the shin at the front of the lower leg. Works in opposition with the gastrocnemius to dorsi flex the foot. It is used to lift the foot off the pedals when driving.
- **Soleus** – lies below and underneath the gastrocnemius in the calf assisting the plantar flexion of the foot.

Functions of the muscular system

Having a basic knowledge of the structure, position and action of the skeletal muscles helps us to understand the functions of the muscular system.

The muscular system has three main functions which are movement, posture and production of heat.

Movement

All types of muscles are responsible for some kind of movement.

- The cardiac muscular tissue causes the movement of the *heartbeat*.
- The visceral muscular tissue of the internal organs moves nutrients and waste through the digestive and renal system. This type of muscular action is known as **peristalsis.**
- The skeletal muscular tissue moves bones at joints known as **isotonic** movement. Muscles can also perform static contractions in which only the muscle moves – this is known as **isometric** movement.

Posture

Skeletal muscles are responsible for maintaining the body in an upright position and in order to be able to do this, muscles contain some fibres that sustain a certain amount

Tip

Round shoulders is a postural problem, which happens as a result of lack of balance between muscles of the chest and back. The chest muscles shorten and the back muscles lengthen. Massage and exercise can help restore balance by relaxing tight muscles and stimulating weak muscles.

Tip

Muscles feel softer when they are warm and are more responsive to massage and exercise than when they are cold.

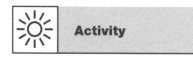

Activity

Did you get it right?

Agonist = gastrocnemeus
Synergist = soleus
Antogonist = tibialis
            anterior

of contraction. This is known as **muscle tone**. If muscle tone is lost completely then the body collapses – a situation which occurs when a person loses consciousness.

- *Good posture* relies on balanced muscle tone in postural muscles.
- Bad posture results in **muscle fatigue** – muscles start to tire and ache as lactic acid builds up.

Production of heat

Active muscles generate large amounts of heat, which is taken to other parts of the body by the blood to help to maintain body temperature.

- If body temperature increases during exercise, vasodilation of blood vessels in the skin and sweating takes place to cool the body (Chapter 2).
- When body temperature drops below a certain level, an involuntary action causes the body to shiver which involves rapid contractions that produce shaking rather than co-ordinated movements. This action generates heat helping to raise the body temperature to within its normal range.
- Muscles also respond to external temperatures becoming more relaxed when warm and tighter when cold.

Common conditions

An A–Z of common conditions affecting the muscular system

- ATROPHY – wasting of muscle tissue.
- CRAMP – a sudden involuntary contraction of a muscle causing acute pain.
- FATIGUE – build up of lactic acid in the muscle causing loss of use.
- FIBROSITIS – inflammation of the muscle fibres.
- MUSCULAR DYSTROPHIES – inherited diseases resulting in the collapse of muscle leading to loss of function.
- MYALGIA – muscle pain.
- MYASTHENIA GRAVIS – chronic disease in which muscles are weak and tire easily.
- MYOKYMIA – persistent quivering of muscles.
- MYOMA – tumour composed of muscular tissue.
- MYOSITIS – inflammation of skeletal muscle.
- MYOTONIA – prolonged muscular spasms.
- PARALYSIS – loss of ability to move a part of the body.
- PARESIS – partial or slight paralysis of muscles.
- RUPTURE – tearing of the muscle fascia or tendon.
- SHIN SPLINTS – soreness in the front of the lower leg caused by excess walking up and down steps or a hill.
- SPASM – a sudden involuntary muscle contraction.
- STRAIN – over use of muscles.
- STRESS – excessive muscular tension resulting in tight, painful muscles and restricted movement in joints.
- TENDINITIS – inflammation of tendon and muscle attachments.
- TENNIS ELBOW – inflammation of the tendons that attach the extensor muscles of the forearm at the elbow joint.
- TENOSYNOVITIS – inflammation of a tendon sheath where it passes over a joint.
- TORTICOLLIS – involuntary contraction of the muscles of the neck. Also known as WRYNECK.

System sorter

MUSCULAR SYSTEM

Skeletal

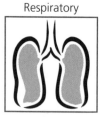

Tendons attach muscles to bones at joints. Bones act as levers and movement is activated as muscles contract.

Respiratory

Breathing in air supplies the muscles with vital oxygen. Oxygen is transported to the muscles by the blood where it reacts with glycogen to produce energy.

Integumentary

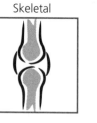

During exercise any excessive heat produced by the muscular system is lost through the skin. Vasodilation causes the skin to redden as heat is lost and the production of sweat causes the body to cool down further.

Circulatory

Circulatory

The blood transports oxygen, glycogen and water to the muscles and transports waste products away to be excreted out of the body.

The digestive system is responsible for processing the food we eat into substances suitable for the muscles to use. Carbohydrates are broken down into glycogen which is stored in the mitochondria within the muscles.

Digestive

Sphincter muscles control the release of urine from the bladder which is under voluntary control. The passing of urine from the kidneys to the bladder is an involuntary action called peristalsis.

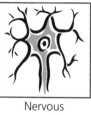

Genito-Urinary

Motor nerves enter the muscle at a motor point. The nerves branch out to supply each muscle fibre with a motor end plate.

Nervous

Hormones produced by the endocrine glands help to regulate the blood flow to the muscles e.g. when the body is under stress the 'fight or flight' syndrome is activated. The hormone adrenaline sends more blood to the muscles to help them cope.

Endocrine

The muscular system consists of three types of muscular tissue, cardiac, visceral and skeletal that are responsible for the involuntary and voluntary movements of the body and its parts.

Holistic harmony

The muscular system plays an important role in maintaining the vital functions of the body, from the cardiac muscle controlling the output of the heart to the smooth, involuntary muscular action of the internal systems of the body, and the voluntary skeletal muscles responsible for the movements made as the human body wanders through life. We rarely give a second thought to many thousands of actions performed by the muscular system during the course of a day in order to carry out the wide variety of tasks we set for ourselves. In order for these tasks to be completed successfully without causing undue strain on the system and body as a whole, the muscular system needs a balance of care.

Fluid

The most important part of the diet for muscles is water as they contain approximately 75 per cent water. Even a small loss in water causes a noticeable drop in muscle strength, power and speed. It is therefore vital that fluid levels within the body are maintained at all times to allow muscles to maintain optimum health. Drinking water before, during and after exercise as well as at regular intervals throughout the day will achieve this.

- Drinking water before exercise avoids dehydration during exercise. Water is stored in the muscle with the glycogen and is released when energy is produced.
- During exercise, the body temperature rises and heat is lost through sweating, potentially causing dehydration. At the same time blood is diverted from the muscles to the skin to cool the body (vasodilation). Drinking water during exercise allows the body to cool down more efficiently through sweating without causing dehydration and at the same time allows the blood to stay at the muscles helping the production of energy.
- Drinking water after exercise helps to 'flush out' the waste products produced in the muscles as a result of energy production e.g. tension produces waste products in the muscles due to muscle fatigue.
- Drinking water regularly throughout the day allows the body to maintain a constant fluid balance vital for homeostasis.

Remember

Remember that fat is stored in the hypodermis and the yellow bone marrow.

Remember

It is advisable to eat at least two hours before exercising in order to gain maximum benefit from the nutrients taken into the body. Eating just prior to exercising will put excessive strain on the body as it will be required to perform two major functions simultaneously, digestion and exercise. The result will be indigestion and fatigue as the body needs time to perform each function well.

   **Activity**

As you read this book, check the tension in your shoulders, neck and face. Are you holding your shoulders high and tight? Is your neck stiff? Are you screwing up your face in concentration? If the answer to these questions is yes, then you need to take a break – relax your shoulders, neck and face and have a glass of water. Limit the amount of time you spend in one position to avoid excessive stiffness.

Nutrition

The important foods necessary for muscle function include **carbohydrates**, **fats** and **vitamins**.

- Carbohydrates are stored as glycogen in the muscles and the liver and are used to create energy. Examples of carbohydrates include: pasta, rice, fruit, beans and lentils.

- Fats provide a back up source of energy and nuts, seeds and their oils provide the most beneficial source of fats.

- Vitamins A, C and E are all antioxidants which help the muscles to use oxygen and detoxify the free radicals, the by-product of energy production. In addition, the B group of vitamins are all essential for the production of energy and can be found in foods such as watercress, mushrooms and tuna.

Rest

In order for muscles to function well, they require adequate resting time to compensate for the levels of activity they undertake. Rest allows the body to repay the oxygen debt associated with excessive muscle use and fatigue, allowing time for the body to rid itself of unwanted waste products which build up as a result of using energy. Rest also encourages the relaxation of muscles, which reduces the amount of muscle fibres contracting at any one time thus reducing excessive tension in the muscles. Deep sleep provides the ultimate rest, but relaxation can be gained in the form of quiet times during the course of the day when we can make a conscious effort to reduce the amount of excess tension in the muscles. Beauty and holistic treatments contribute greatly to the well-being of the muscular system. Lying on a couch encourages body and mind to relax, and the physical act of massage stimulates blood flow and warms the muscles so they can be gently restored to normal.

Activity

A balance of varied activity in the form of physical exercise plays an important part in maintaining healthy muscles. Exercise provides the body with the means to develop increased levels of power, strength, speed, endurance and flexibility. Once a level of fitness has been achieved it needs to be maintained and therefore regular exercise is essential. It is recommended that twenty minutes of exercise should be taken three times a week to maintain a healthy muscular system. The type of activity should be varied and interesting, with the aim of working the major muscle groups together with the heart and lungs. This will increase oxygen uptake and increase the amount of exercise you can perform before becoming out of breath. Our lives are filled with many labour-saving devices which have the potential to make us less active as a result. Therefore there is an even greater need for regular exercise/activity as our lives become physically easier.

Air

Muscles need a good supply of oxygen to activate energy. The type of breathing performed and the quality of air both play an important role in maintaining this function. During exercise it is important to ensure that efficient breathing takes place allowing time to take a deep breath into the body before allowing the breath out to take place. When performing muscular strength exercises, the breath *in* should take place during relaxation of muscles and the breath *out* during exertion of the muscles i.e. when performing a sit up, breathe *out* as you sit up and *in* when you relax. This will allow the muscular system to perform at optimum levels and help to prevent muscle fatigue. It is amazing how often we forget to breathe at all when exercising or concentrating hard! The body is then unable to perform as well as it could and so care should be taken to ensure that the depth of breathing matches the level of activity.

Age

Muscles weaken with age and lack of activity. These factors combined with the gradual slowing down process of the body in general contribute to the formation of lines, wrinkles and dropped contours as weakened muscles are less able to resist the pull of gravity. Regular exercise helps to keep muscles fit and healthy which in turn has a 'knock on' effect on the integumentary (Chapter 2) and skeletal systems (Chapter 3) by stimulating the circulation and helping cellular function. This in turn improves the external appearance of the body as well as its functions. All of the other systems of the body benefit from this use of the muscular system helping to ensure the longer survival of the human body.

Colour

The muscular system relies on the skeletal system (Chapter 3) to provide it with levers in the form of bones for

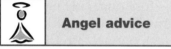

attachment and joints to allow movement to take place. It also relies on instruction from the nervous system (Chapter 9) and energy in the form of oxygen and glycogen from the respiratory (Chapter 5) and digestive systems (Chapter 7) to initiate the movements. The position of the chakras along the axial skeleton together with their associated colours has a collective effect on the well-being of the muscles. Visualising the spectrum of colours will help to energise the body as a whole and as up to 50 per cent of the body may be made up of muscular tissue, this will have an inevitable 'knock on' effect on the system. Individual colours can also have more specific actions e.g. the stimulating effect of red, soothing effect of blue etc.

Awareness

The muscular system supplies us with a unique method of non-verbal communication – body language. Facial muscles are responsible for the formation of facial expressions, which mirror our thoughts and convey a variety of feelings to those around us. Muscles of the body enable us to choose the way in which we control our movements, allowing us to present ourselves in different ways depending on each situation. Body language usually takes the form of an intuitive response and the correct interpretation tells us more about a person than their words alone can ever do.

Special care

The primary aim of the muscular system is to maintain muscle tone, thus providing a starting point for movement, which in turn assists body temperature. Posture is the term used to describe the alignment of the body and plays an important role in ensuring the safe and healthy working of the muscular system. Correct body posture contributes to the overall well-being of the whole body in the following ways:

- It allows full, deep and unrestricted breathing to take place.
- Digestion can function efficiently as good posture prevents the digestive organs from being compressed in the abdomen.
- Postural problems are avoided due to body weight being evenly distributed.
- Good posture eliminates figure faults resulting in a more flattering body image.

Treatment tracker

MUSCULAR SYSTEM

Make up

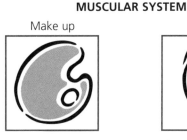

Make up can be used to enhance body shape eg to highlight cleavage.

Facials

Facial massage relieves the everyday tension that builds up in the tiny facial muscles.

Nail care

Hand and foot massage stimulates the extensor and flexor muscles aiding movement.

Hair removal

Many athletes have the excess terminal hairs on their body removed to increase their speed. e.g. swimmers, runners etc.

Massage speeds up the removal of lactic acid and carbon dioxide from the muscles relieving the ache associated with muscle fatigue.

Massage

EMS works weak muscles by stimulating the motor point to produce a contraction. Regular treatment will improve muscle tone.

Working over the reflex points for the joints of the body relieves tension and can help conditions such as frozen shoulder.

A blend of camphor, sage and lemongrass mixed with a carrier oil will be of benefit to tense, tight muscles.

Electrical

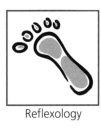

Reflexology

Aromatherapy

The skeletal muscles of the body receive many benefits from beauty and holistic treatments helping to stimulate blood flow and in turn improve energy production and movement.

Knowledge review – Muscular system

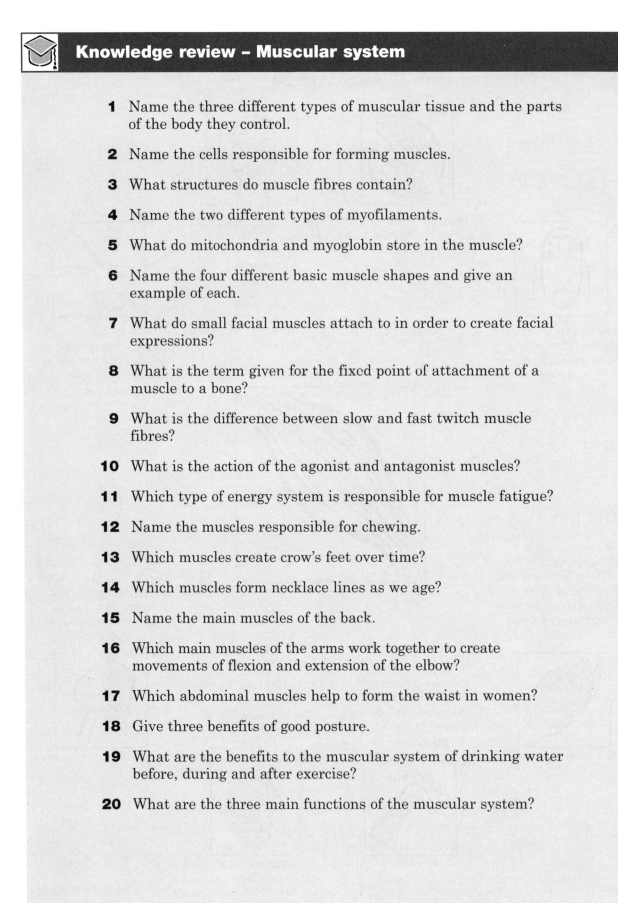

1   Name the three different types of muscular tissue and the parts of the body they control.

2   Name the cells responsible for forming muscles.

3   What structures do muscle fibres contain?

4   Name the two different types of myofilaments.

5   What do mitochondria and myoglobin store in the muscle?

6   Name the four different basic muscle shapes and give an example of each.

7   What do small facial muscles attach to in order to create facial expressions?

8   What is the term given for the fixed point of attachment of a muscle to a bone?

9   What is the difference between slow and fast twitch muscle fibres?

10   What is the action of the agonist and antagonist muscles?

11   Which type of energy system is responsible for muscle fatigue?

12   Name the muscles responsible for chewing.

13   Which muscles create crow's feet over time?

14   Which muscles form necklace lines as we age?

15   Name the main muscles of the back.

16   Which main muscles of the arms work together to create movements of flexion and extension of the elbow?

17   Which abdominal muscles help to form the waist in women?

18   Give three benefits of good posture.

19   What are the benefits to the muscular system of drinking water before, during and after exercise?

20   What are the three main functions of the muscular system?

The respiratory system

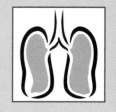

5

Learning objectives

After reading this chapter you should be able to:

- **Recognise the parts that make up the respiratory system**

- **Identify the different processes involved with respiration**

- **Understand the functions of the respiratory system**

- **Be aware of the factors that affect the well-being of the respiratory system**

- **Appreciate the ways in which the respiratory system works with the other systems of the body to maintain homeostasis.**

The next few chapters are going to take us through the systems involved with the input and output functions of the body. The main input vital to maintaining life is **oxygen**, which we breathe in from the air around us. As a result of using this oxygen, the body produces **carbon dioxide**, most of which is released from the body as output when we breathe out. This process is known as **respiration** and the system responsible for ensuring that this process works efficiently is **THE RESPIRATORY SYSTEM.** The respiratory system lies in the upper most section of the body starting at the nose extending down to the chest area below the breasts. It consists of external openings into the body i.e. the *nose* and *mouth* together with a network of *air passageways* leading to the **lungs**. It is separated from the lower section of the body by the **diaphragm**. The respiratory system is closely linked to the muscular system (Chapter 4) and the skeletal system (Chapter 3) for protection and support, the circulatory systems (Chapter 6) as it is the blood that is responsible for the transportation of oxygen and carbon dioxide, and the nervous system (Chapter 9) which controls the rate and depth of breathing. In turn the respiratory system is responsible for the well-being of every body system because of their continuous need for oxygen. For this reason, oxygen is often referred to as the 'life force'.

Science scene

Structure of the nose, pharynx, trachea, bronchi and lungs

The structures of the respiratory system are designed to facilitate the most important process of the body – that of breathing. Air from the atmosphere around us is inhaled into the body; a process follows which allows the exchange

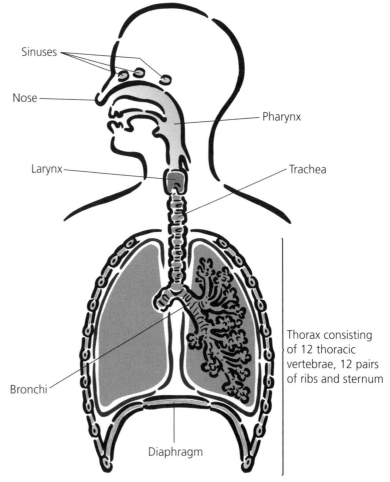

The respiratory system

of oxygen and carbon dioxide to take place before air is exhaled out of the body. This process is repeated many thousands of times throughout the course of a day and night and is vital to the life of every individual cell, tissue, organ and system of the body.

The respiratory system can be divided up into two main groups – the *upper* and *lower respiratory tracts*:

1. Upper respiratory tract consisting of:
   - Nose
   - Sinuses
   - Pharynx
   - Larynx

2 Lower respiratory tract consisting of:
   - Trachea
   - Bronchi
   - Lungs

The **thorax** protects the lower respiratory tract and consists of:

- 12 pairs of **ribs** creating a cage like structure
- 12 **thoracic vertebrae** providing connection at the back
- The **sternum** providing connection at the front

Tip

Chapter 3 The Skeletal System provides detailed information of the thorax.

Structures of the upper respiratory tract

Eustachian tube

Sinuses

Nasal cavity

Nasopharynx

Mouth

Tonsil

Oropharynx

Laryngopharynx

Epiglottis

Trachea

Oesophagus

Upper respiratory tract

Tip

Chapter 3 The skeletal system provides detailed information on the bones and cartilage.

Fascinating Fact

Irritation of the nose results in the reflex action of sneezing which forcibly expels the unwanted particles causing the irritation!

Tip

The mouth also forms an entry and exit point for air and is discussed in more detail in Chapter 7 The digestive system.

Tip

Massage can help to drain the sinuses restoring normal function and the client may need to blow their nose as a result!

Nose

The nose is the main point of entry for air coming into the body and the main point of exit for air leaving the body. The nose is made up of:

- The **nasal** bones forming the bridge of the nose
- The **turbinate** bones forming the sides of the nose
- **Hyaline cartilage** forming the flexible shape of the front of the nose.

The nostrils form two separate openings into the *nasal cavity*; these openings are divided by a thin wall of cartilage known as the **septum**. The nasal cavity is lined with *ciliated mucous membrane* forming layers of cells with hairs called **cilia** which filter air as it enters. The mucus is produced by goblet cells and is slimy and thick trapping any minute particles in the air as it enters the nose.

Sinuses

The sinuses are air spaces in the bones of the frontal, ethmoid, sphenoid and maxillae bones which all open onto the nasal cavity. Mucous membrane lines the sinuses and is continuous with the nasal cavity. Painful headaches can result from mucus blockages in the sinuses.

Pharynx

The nasal cavity leads to the pharynx (back of the throat) which is also lined with a ciliated mucous membrane. It consists of muscular and fibrous tissue and can be divided into three sections:

1. *Nasopharynx* or nasal part of the pharynx provides a passageway for air when breathing through the nose. It contains an opening to each ear called the **eustachian tubes** containing mucus. Throat infections can easily spread to the ear via the eustachian tubes. The **adenoids** are found in this section of the pharynx. Made of lymphatic tissue, they contribute to the immune functions of the body by filtering harmful substances from the air.

2. *Oropharynx* or oral part of the pharynx is the passageway for air and food from the mouth. It contains the **tonsils** which like the adenoids, also contribute to the immune functions of the body.

3. *Laryngopharynx* is the passageway for food to enter the oesophagus, which is the first part of the digestive system leading to the stomach.

Larynx

The pharynx leads to the larynx (upper throat) which is the next passageway for the air entering the body. The larynx continues the filtering action of the airways. It contains cartilage forming the **vocal cords** and it is for this reason that the larynx is also known as the **voice box**. Cartilage also forms a lid-like structure called the **epiglottis** which overhangs the entrance to the larynx. The epiglottis is responsible for preventing food from entering the airways when we swallow.

Structures of the lower respiratory tract

Trachea

The **trachea** or windpipe, as it is more commonly known, leads on from the larynx and extends down into the thorax. Filtering of the air continues to take place in the trachea through a mucous lining. The trachea is made up of C-shaped hyaline cartilage at the front, which is connected to make a circle by visceral muscle, and connective tissue at the back. This semi-solid structure prevents the trachea from collapsing, blocking the passage of air. The trachea extends about 12 centimetres down into the thorax

Remember

Ears are not considered to be a part of the respiratory system but as they have a connection with the upper respiratory tract, it is useful to have an awareness of their links with the system.

Fascinating Fact

The size of the larynx is the same in children of both sexes until they reach puberty when the vocal cords grow larger in males contributing to the structure of the 'Adam's apple' and the deepening of the voice.

Activity

If you touch the centre of your neck you will feel the hard cartilage of the trachea and be able to track it down towards the sternum. It is an area that needs lighter pressure when performing neck massage to avoid discomfort.

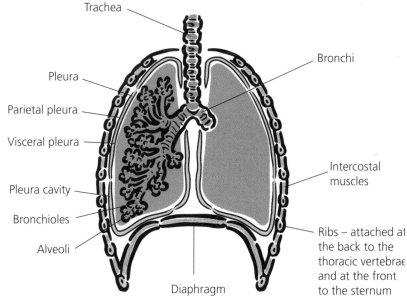

Lower respiratory tract

Remember

The left lung is the smaller of the two to allow room for the position of the heart within the thorax.

before dividing into two sections – the left and the right **bronchi.**

Bronchi

The bronchi are passages similar in structure to the trachea. They provide a passageway for the air into the left and right lungs. The left bronchus is narrower and shorter than the right and divides into two branches as it enters each section or lobe of the left lung. The right bronchus divides into three branches entering the three lobes of the larger right lung.

The bronchi continue the filtering process of the air through their mucous lining.

Lungs

The lungs are soft, spongy, balloon-like structures situated on either side of the heart within the thorax. Each lung contains a bronchus that subdivides to enter each lobe.

Within the lobes the bronchi subdivide further forming smaller tubes called **bronchioles.** These small tubes have lost their cartilage structure and contain only visceral muscular tissue making them soft. **Alveoli** form at the end of the bronchiole tubes. These are tiny air sacs which are bathed in blood and surrounded by a network of tiny blood capillaries. The interchange of the vital gases oxygen and carbon dioxide takes place within the blood of the alveoli and the surrounding capillaries.

The lungs have an outer protective covering or membrane known as the **pleura.** The pleura is made up of two layers:

- **Visceral** inner layer which is attached to the lungs
- **Parietal** outer layer which is attached to the ribs and the diaphragm.

The visceral and parietal layers of the pleura are divided by the *pleural cavity* which contains a lubricating fluid, and allows movement between the two layers as breathing takes place.

Functions of the respiratory system

Respiration is the process of the exchange of oxygen and carbon dioxide. Oxygen needs to be breathed into the body, transported to the cells by the blood (Chapter 6) in order for

Fascinating Fact

The reverse action takes place in trees and plants. In sunlight, they take in carbon dioxide and release oxygen. This process is called **photosynthesis** and is why the air in the country is so much better for us than that in the cities.

Activity

The sense of smell encourages us to move away from air that has an unpleasant odour so that we can ensure that the body breathes in fresh air whenever possible.

nutrients from the digestive system (Chapter 7) to be oxidised i.e. broken down, ATP to be produced in the muscles (Chapter 4) and ultimately for energy to be released. All cells within the body need a constant supply of oxygen to keep them alive. Carbon dioxide is produced as a result of the use of oxygen within the body and needs to be transported by the blood from the cells to the lungs to be breathed out so that the process can begin again. We can survive for a few weeks without food, a few days without water but only a few minutes without oxygen!

Respiration involves five individual processes: breathing, external respiration, transportation, internal respiration and cellular respiration.

Breathing

Air is brought into the body through the nose or mouth. Breathing air in through the nose is a more efficient process than breathing air in through the mouth because within the nasal cavity the air is:

- Filtered by the cilia inside the nasal cavity to remove unwanted substances; these are either eliminated from the nose when we sneeze or blow our nose, or pushed down towards the laryngopharynx with the mucus where it is then swallowed.
- Warmed by the blood in the nose adjusting the temperature of the air to that of the body.

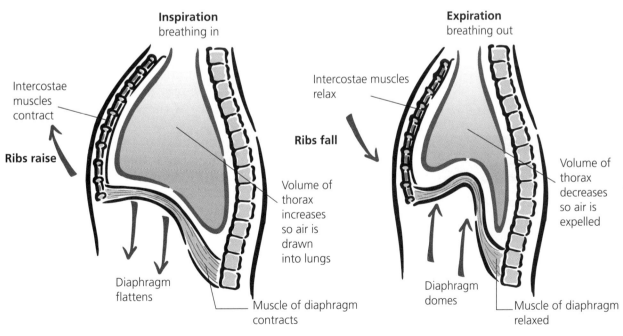

Respiration – inspiration and expiration

- Moistened as water evaporates from the mucous lining.
- Smelt by the sensory nerves in the nose passing the information to the brain, which in turn identifies the smell.

Breathing can be defined as the movement of air in and out of the lungs by inspiration or inhalation and expiration or exhalation:

Inspiration – breathing *in*.

- Diaphragm contracts pushing abdominal cavity down
- Intercostal muscles contract
- Ribs are pulled up and out
- Thoracic cavity enlarges
- The pressure in the lungs decreases
- The pressure of the incoming air increases
- Air rushes into the lungs
- Lungs expand as they fill with air.

Expiration – breathing *out*.

- Diaphragm relaxes and returns to dome shape
- Intercostal muscles relax
- Ribs return to normal position
- Thoracic cavity returns to original shape
- The pressure in the lungs increases
- The pressure of the air outside decreases
- The air is able to flow out of the lungs
- Elastic recoil of lungs helps to force air out
- Contraction of the abdominal muscles will aid forced expiration by pushing the organs of the lower trunk up so increasing the force of the air out of the body.

Fascinating Fact

We breathe in and out over 1,000 times per hour and over 25,000 times every day!

After exhaling there is a slight pause before the next breath when the pressure inside the lungs equals the pressure of the outside air; this is known as a *state of equilibrium*.

Breathing is controlled by the **nervous system** (Chapter 9) and happens without conscious thought or effort. The rate of breathing changes as changes occur within the body e.g. when we need to run for the bus, our rate of breathing increases to ensure that enough oxygen gets to the muscles to activate the energy needed for such a task. When we have caught the bus and sat down, our rate of breathing slows down as the muscles now need less oxygen.

External respiration

The exchange of oxygen from the air with carbon dioxide in the blood takes place within the alveoli in the lungs. This exchange of gases occurs because of a difference in pressure and concentration levels in the alveoli and the blood capillaries in the lungs:

- As the air enters the alveoli with oxygen, it is under greater pressure than the blood in the surrounding capillaries. As a result, the oxygen can pass easily through into the blood raising its concentration levels. When the pressure becomes equal this process, which is known as **diffusion**, stops.

- The carbon dioxide in the blood, which has come from the cells of the body, is under greater pressure than the air in the alveoli, which contains small amounts of carbon dioxide with the inspired air. As a result, the carbon dioxide in the blood can diffuse through from the capillaries into the alveoli raising their concentration levels.

Transportation

The transportation of oxygen and carbon dioxide is performed by the **pulmonary circulation**:

- Because of the exchange of gases in the alveoli, the blood can then transport the oxygen to the heart via blood vessels called the *pulmonary veins*, where it is pumped around the body to all the cells to be used and replaced with carbon dioxide.

- The blood then transports the carbon dioxide back from the cells to the heart where it is taken to the lungs via the *pulmonary arteries*, to be released with the expired air.

Internal respiration

Transportation ensures that the **oxygenated** blood (blood with a high concentration of oxygen) gets to the cells where the exchange of oxygen and carbon dioxide takes place by the same method of diffusion:

- The pressure of oxygen in the blood entering the cells is greater, allowing the transfer of oxygen to take place.

- The pressure in the blood leaving the cells is lower, allowing the transfer of carbon dioxide to take place.

The oxygen is replaced with carbon dioxide and the blood becomes **deoxygenated** and begins the whole cycle again.

Remember

A certain amount of air remains in the lungs at all times. This air contains a balance of oxygen and carbon dioxide necessary to maintain homeostasis (homeo = same; stasis = state) and is controlled by the brain. The carbon dioxide is used by the body to maintain pH balance. Breathing deeply when the body does not require it produces oxygen in excess of body requirements and lowers the carbon dioxide levels, so altering the pH balance. This results in the blood vessels to the brain constricting, reducing the flow of blood and often producing a panic attack. Breathing into a paper bag for a few moments increases the carbon dioxide levels creating balance once again.

Cellular respiration

Cellular respiration refers to the utilisation of oxygen in the cells and production of carbon dioxide. The individual cells use the oxygen to form energy and as a result produce carbon dioxide.

It is important to appreciate that every living cell is dependent on the act of breathing for survival and care should be taken to ensure that the rate and depth of breathing matches the needs of the body. Although this is controlled by the autonomic nervous system (Chapter 9), everyday factors such as stress and poor posture can put excessive strain on the respiratory system resulting in poor breathing techniques. This in turn affects the performance of the cells, tissues, organs and systems of the body.

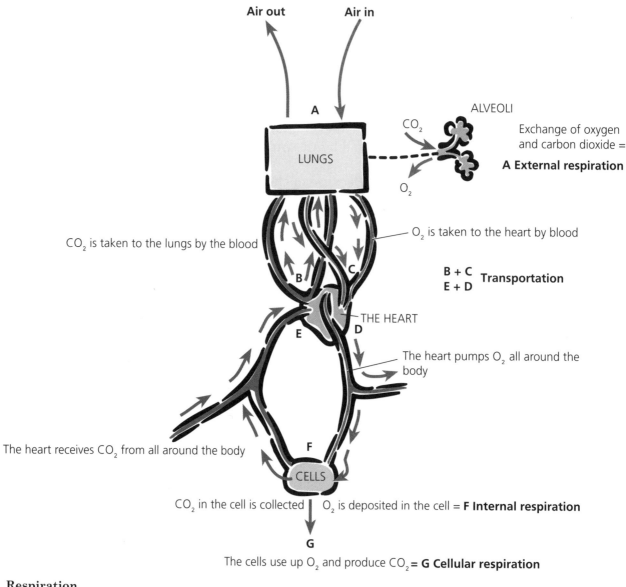

Respiration

Therapists should pay attention to the process of their own breathing whilst performing treatments, in the same way as attention should be paid to the breathing techniques of their clients. A therapist's breathing will increase with the level of physical activity as the client's breathing decreases with the level of relaxation.

Common conditions

An A–Z of common conditions affecting the respiratory system

- ADENOID enlarged – can block the opening of the eustachian tube and/or obstruct air flow from the nose to the throat.
- ASTHMA – difficulty in breathing caused by narrowing of the airways. It may be triggered off by external factors, known as extrinsic asthma, or internal factors, known as intrinsic asthma.
- BRONCHITIS – inflammation of the lining of the bronchi.
- CLEFT PALATE – a deformity of the palate. This is often accompanied by a hare-lip.
- COMMON COLD – contagious viral infection resulting in a sore throat and runny nose. Usually lasts for 2–7 days with full recovery taking up to three weeks.
- CROUP – viral infection that affects children. Characterised by a hoarse, barking cough and a fever.
- EMPHYSEMA – inflammation of the alveoli in the lungs causing blood flow through the lungs to slow down. It is usually associated with bronchitis and/or old age.
- GLANDULAR FEVER – a viral infection which is most common in the 15–22 year old age group. It is characterised by a persistent sore throat and/or tonsillitis.
- HAYFEVER – caused by an allergy to pollen. Hayfever commonly affects the nose, eyes and sinuses as the pollen irritates these areas causing excessive sneezing, watery eyes and a build-up of mucus. The airways may also be affected resulting in wheezing as breathing becomes more difficult.

- HYPERVENTILATION – rapid deep breathing commonly associated with stress.
- LARYNGITIS – inflammation of the larynx producing hoarseness and/or loss of voice. There are two forms: acute which develops quickly and is short lived, and chronic, which is recurring.
- LUNG CANCER – a life threatening, malignant growth (tumour) in the lungs.
- NASAL POLYP – harmless extensions of the mucous lining within the nasal cavity containing fluid causing an obstruction in air flow.
- PHARYNGITIS – inflammation of the pharynx resulting in a sore throat and may be either acute or chronic. Acute pharyngitis is very common and clears up after a week or so. Chronic pharyngitis is longer lasting and can result from smoking.
- PLEURISY – inflammation of the pleura surrounding the lungs and usually occurs as a complication of other disorders.
- PNEUMONIA – inflammation of the lungs from either a bacterial or viral infection resulting in chest pain, dry cough, fever etc. Bacterial pneumonia tends to last longer.
- PNEUMOTHORAX – collapsed lung. Can be caused by rupture to the lungs.
- RHINITIS – inflammation of the mucous lining of the nasal cavity causing a blocked, runny and stuffy nose.
- SINUSITIS – inflammation of the mucous lining of the sinuses causing a blockage which can be very painful, swollen and sore. It can be acute, usually accompanying a cold, or chronic when there are recurrent blockages.
- STRESS – excessive stress causes the autonomic nervous system to activate the release of the hormone adrenalin. This causes the breathing rate to increase.
- TONSILLITIS – inflammation of the tonsils resulting in soreness in the throat. It is more common in children.
- TUBERCULOSIS (TB) – infectious disease involving the formation of nodules in body tissue. The most common sites are the lungs. Immunisation is available.

System sorter

RESPIRATORY SYSTEM

Skeletal

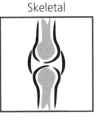

The bones of the thorax – 12 thoracic vertebrae, 12 pairs of ribs and the sternum – protect the organs of respiration.

Muscular

The intercostal muscles and the diaphragm aid in the process of breathing by contracting to enlarge the space to let air into the lungs.

Integumentary

The skin relies on the oxygen processed by the respiratory system for the renewal of skin, hair and nail cells.

Circulatory

Blood transports oxygen breathed into the lungs around the body to each cell. It also transports carbon dioxide from the cells to the lungs to be released out of the body as we breathe out.

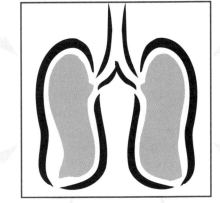

The mouth and pharynx link the digestive and respiratory systems and the muscular co-ordination in the throat separates them.

Digestive

The kidneys compensate for water lost through breathing by monitoring fluid levels in the body.

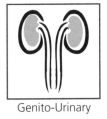

Genito-Urinary

The nervous system is able to 'pick up' on the body's need for oxygen and activate the diaphragm to contract which ensures breathing takes place.

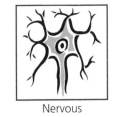

Nervous

The hormone adrenalin produced by the adrenal glands is released into the bloodstream and changes the rate of breathing during times of stress.

Endocrine

The respiratory system is responsible for a complex set of actions vital to maintaining human life including:

- Breathing – inhalation and expiration
- External respiration – the interchange of oxygen and carbon dioxide in the lungs
- Transportation – of oxygen to the cells from the lungs and carbon dioxide from the cells *to* the lungs.
- Internal respiration – the interchange of oxygen and carbon dioxide in the cells
- Cellular respiration – the use of oxygen by the cells.

Holistic harmony

The respiratory system plays a vital role in keeping the body alive. It provides the entry point for oxygen, the 'life force' of the body and the exit point for unwanted carbon dioxide. This process services each and every cell, tissue, organ and system and is responsible in part for every activity the body performs. Efficient breathing therefore contributes to the efficient performance of the body as a whole. However, in order for this vital service to be maintained, the respiratory system requires a certain amount of care.

Fluid

We are already aware that the body is made up of large amounts of water. A small percentage of this is lost during the day through breathing. To preserve this water, it is better to breathe out through the nose. This ensures that much of the water is kept within the body as it is trapped in the nose by the ciliated lining. More water is lost in a breath out through the mouth and can result in a drying out of the mouth. Transportation of oxygen and carbon dioxide is improved when the water content of the body is good. The blood is formed into a liquid substance because of the amount of water present in the body. Therefore, the transportation of oxygen and carbon dioxide by the blood is much more efficient when water levels are high. Dehydration causes the blood to thicken and transportation becomes less efficient having a 'knock on' effect on the delivery of oxygen to the cells and carbon dioxide away from the cells. The body is able to use fresh, clean water better and more quickly than any other forms of fluid and it is for this reason that we should pay attention to our daily intake and try to follow the recommended guideline of approximately one and a half litres of water per day.

Nutrition

Lung cancer is one of the biggest causes of death in the West and there is increasing evidence that the food we eat is a contributory factor. Progress is currently being made in nutritional therapy as a form of both prevention and treatment of cancer, with the following findings:

- Vitamins A, C and E help counteract the debilitating effects of free radicals.
- Selenium has antioxidant properties and is recommended for people who have a history of

Fascinating Fact

When we haven't been breathing deeply enough we yawn. This provides the body with an extra boost of oxygen into the lungs with the in breath and maximum loss of carbon dioxide with the out breath, thus improving circulation and making us feel more awake.

cancer in their family and/or who smoke. Good sources of selenium are tuna, mushrooms and cottage cheese.

We are what we eat, so it is important to take responsibility for what we are by making the most of our diet. This is a philosophy that can be used in the after care advice given at the end of treatments to encourage a client to learn to help their own body by taking note of their own actions. We all need a little help in the quest for balance and harmony.

Rest

When we sleep our breathing slows down and matches the intake of oxygen with the needs of the body. At other times our breathing rate fluctuates, so times of rest are needed in order to restore the oxygen levels in the body; slow, deep breathing concentrating on a good breath in and a good breath out will aid relaxation, after experiencing a stressful situation or an intense exercise session, and will restore the balance. The rate of breathing is controlled by the nervous system (Chapter 9) – as oxygen levels dip the diaphragm receives a message from the brain causing it to contract and inhalation takes place.

When the body needs oxygen quickly, we tend to breathe in through the mouth which provides a quicker route in.

When we want to force air out of the body quickly, we tend to breathe out through the mouth.

Activity

In the previous chapter we recognised that good posture contributes to good breathing so exercises which concentrate on the postural muscles are of extreme benefit. This helps to keep the chest area wide and open as opposed to closed and slumped, helping the intake and output of air through the system. In addition, exercise strengthens the muscles associated with breathing, i.e. intercostals and diaphragm, which allows breathing to become efficient. Activity that increases the heart rate is beneficial as it strengthens the lungs, increasing their capacity for air. Walking instead of driving, using the stairs instead of the lift, housework and gardening are all good examples of the type of everyday activity that is of direct benefit to the respiratory system. The living body never stops breathing – it just slows down and speeds up depending on the level of activity.

Fascinating Fact

The lungs of a baby are quite pink and in an adult they are quite grey and can even be black! Years of breathing in dirty air can cause this change!

Activity

Think about the stress that sitting at a school desk puts on a child's posture as he or she 'learns' to slouch. This continues into teenage and adult years as the body responds to increased workloads as well as emotional stress. It is no wonder that the majority of people do not breathe as well as they should!

Angel advice

Methods such as the Alexander Technique aim to re-balance posture, breathing and well-being.

Air

The type of air breathed into the body will affect the functioning of the respiratory system and in turn the body as a whole. The air we breathe contains other gases in addition to oxygen some of which are harmless and some of which are extremely harmful. The clean air associated with the countryside is rich in oxygen from the photosynthesis action of trees and plants. The air associated with a busy city, on the other hand, has a higher content of harmful gases such as car fumes, factory emissions etc. The body has to cope with the additional gases it breathes in and this puts extra strain on the system. Smoking contributes to the added pressure put on the system and is the cause of many disorders associated with the respiratory system which can be life threatening. The ciliated linings of the respiratory tract help to rid the body of the toxins associated with polluted air in the short term but cannot cope in the long term and this results in the onset of disease.

Age

A baby gradually learns to use its body intuitively with a natural sense of co-ordination. As we age, the body adapts to its environment both physically and emotionally and this is reflected in the way in which we breathe. The young, uninhibited body works in such a way that allows the respiratory system to function at optimum levels, resulting in unrestricted breathing at all times. As we become affected by stress, our body tenses and this affects the level of breathing that takes place.

Colour

In previous chapters we have explored the benefits of visualising colours to assist the well-being of the body and its parts. We can also actively incorporate the use of colour when breathing to further enhance the therapeutic effects of the different colours as well as encourage efficient breathing techniques. This can be carried out in two ways, either to calm the whole body or a specific part of the body depending on the needs of the person and the choice of colour:

- A generally stressed person would benefit from focusing on visualising the complete rainbow with their inward breath to help to re-balance the body. The outward breath will signify the physical and psychological release of the stress-causing factors.

Stress activates the release of the 'fight or flight' hormone adrenaline (Chapter 10) into the blood stream, which changes the rate of breathing making it shallow and rapid. This type of breathing is suitable in the short term but detrimental to the body if sustained over long periods.

Angel advice

It is important to respect the invisible barrier between people when dealing with clients ensuring that they feel 'safe' before you invade their 'space'.

Activity

How many times do you find that you have stopped breathing whilst concentrating on a difficult task? Do you think that this helps or hinders the performance of the task?

- A person with a tension headache will gain a sense of calmness from focusing on the visualisation of the colour violet with the inward breath releasing negative thoughts with the outward breath. Violet is the colour associated with the head generally, in particular with the brain and higher thinking. This activity will help to restore a clearer mind and a calmer body.

The mere thought of taking in colour as we breath has the effect of slowing down our breathing as we focus on something other than the thing which has caused us stress. This automatically calms the system allowing more oxygen into the body and carbon dioxide out restoring the equilibrium of the body.

Awareness

Each time we breathe in we are taking into our body millions of molecules. These molecules include the vital ones of oxygen, as well as carbon dioxide, bacteria and viruses that the body's immune system has to process, and also the air other people have breathed out! These factors bind us to everything that is around us thus forming a connection which enables us to respond accordingly to the environment. In this way we could imagine that our breath is an open passageway linking us at some level with the people, animals, plants etc. around us. So we could say that breathing provides the link internally between body systems and externally between all living things! The invisible barrier created between people is an indication of the links we choose to make as we intuitively move away from those people who upset us and move closer to those we care for.

Special care

Should be taken in the way in which we breathe, as poor breathing results in a reduction in the amount of oxygen available for each cell, causing the body to experience many problems. Examples of this include: muscle cramps, headaches, depression, anxiety, chest pains, tiredness etc. An awareness of the way in which we should breathe is crucial in avoiding such problems.

- **Lateral costal** – normal breathing, where the lungs take in enough oxygen to accommodate everyday activities. This type of breathing is associated with the **aerobic** energy system and involves filling the upper two lobes of the lungs with air.

- **Apical** – shallow and rapid breathing is used when the body wants to get maximum amounts of oxygen to the muscles. Examples of those times include exercise, childbirth, stress, fear etc. This type of breathing is associated with the **anaerobic** energy system and results in oxygen debt and muscle fatigue if energy requirements exceed oxygen intake. Air is taken into the upper top lobes of the lungs only.

- **Diaphragmatic** – the deep breathing associated with relaxation which repays any oxygen debt accumulated through apical breathing and aims to fill all the lobes of the lungs.

Angel advice

Free your mind and body with your breath. The art of living is reflected in the art of breathing and allows a flow of positivity into the body and negativity out of the body

The skills associated with correct breathing can be learned and techniques such as yoga and tai chi have a strong focus on the art of breathing.

Breathing techniques should accompany beauty and holistic treatments where appropriate as they are of benefit to both the therapist and the client, and help to clear the mind and energise the body.

- Starting the treatment with some deep breathing exercises will help to calm a stressed client and prepare them for treatment.

- Completing a treatment with some breathing exercises is a way of encouraging long-term benefits by making the client aware of the links between their breathing and their stress levels.

Finally, every therapist should be aware of the benefits of practising good breathing on the performance of their treatment.

- Taking a few deep breaths before the start of a treatment helps the therapist to focus on the client and their needs, clearing the mind of other things.

- Paying attention to breathing during treatment will allow the body to react to changes as the physical demands of performing the treatment change, requiring different levels of activity from the therapist together with differing amounts of oxygen.

- A few deep breaths at the end of the treatment allow the mind to clear in preparation for the next client etc.

Breathing is an underestimated task and one which is taken for granted in everyday life. Special care should be taken to ensure that the respiratory system is able to perform its functions freely and efficiently without undue strain or discomfort.

Treatment tracker

RESPIRATORY SYSTEM

Nail care

Make up

Facials

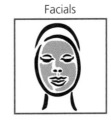

The physical outward appearance of an individual affects confidence levels which in turn affects breathing. Painted nails and a well made-up face increases confidence levels improving breathing and well-being.

Facial massage can be used to drain the excess mucus associated with sinus problems.

Hair removal

The stress associated with excessive hair growth changes the rate of breathing. Hair removal increases self esteem having a 'knock on' effect on the respiratory system through better breathing.

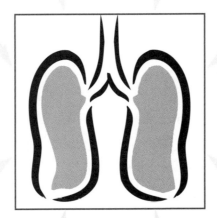

Massage induces relaxation which automatically creates balanced breathing and improves respiration.

Massage

Relaxation movements used at the start of a reflexology treatment help to balance the breathing which has a clearing effect on the mind and energises the body.

Eucalyptus oil is a valuable decongestant for respiratory tract infections helping to clear the airways.

Electrical face and body treatments help to improve muscle tone which in turn aids posture. Good posture contributes to good breathing.

Electrical

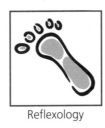

Reflexology

Aromatherapy

Beauty and holistic treatments provide the respiratory system with the perfect means to relaxation helping to re-balance and re-educate.

Knowledge review – Respiratory system

1 Name the parts that make up the upper respiratory tract.

2 Name the parts that make up the lower respiratory tract.

3 As air enters the body which structures help to filter it?

4 Explain why is it more effective to breath in through the nose.

5 What are the three parts of the pharynx called?

6 What is the name of the tube that connects the ear with the nasal cavity?

7 Which structure contains the voice box?

8 Which part of the respiratory system prevents fluid and food from entering the lower respiratory tract when we swallow?

9 What is the trachea commonly known as and where is it situated?

10 Which lung is the smallest and why?

11 Name the inner and outer layers of the pleura.

12 What are the layers of the pleura filled with and why?

13 Name the structures in the lungs where the exchange of oxygen and carbon dioxide take place.

14 Give the technical terms to describe breathing in and out.

15 What are the differences between external, internal and cellular respiration?

16 Why is country air better than city air?

17 How does good posture contribute to good breathing?

18 Which type of breathing is associated with stress?

19 Give three possible problems that may occur as a result of poor breathing.

20 Why do lungs change from being pink in colour to grey or possibly black?

The circulatory systems

6

Learning objectives

After reading this chapter you should be able to:

- **Recognise the structures that make up the circulatory systems**

- **Identify the main blood and lymph vessels**

- **Understand the functions of the circulatory systems**

- **Be aware of the factors that affect the well-being of the circulatory systems**

- **Appreciate the ways in which the circulatory system works with the other systems of the body to maintain homeostasis.**

The body relies on a complex system to ensure that the vital input and output functions of individual cells are maintained. The system responsible for this activity is the **CIRCULATORY SYSTEM.** The circulatory system consists of two complementary systems – the *blood* circulation and the *lymphatic* circulation which work together in providing the body with a transportation system.

The blood circulation is a two-way system:

1. Transporting substances such as oxygen, water and nutrients necessary for survival *to* every cell in the body.

2. Transporting unwanted substances such as carbon dioxide and waste products *away* from every cell to be released out of the body.

The lymphatic system is a one-way system which supports the blood circulation by:

1. Transporting *away* from the cells the unwanted substances that the blood is unable to take.

Science scene

Structure of the circulatory systems

1. The blood circulatory system consists of:
   - Blood
   - The heart
   - Blood vessels

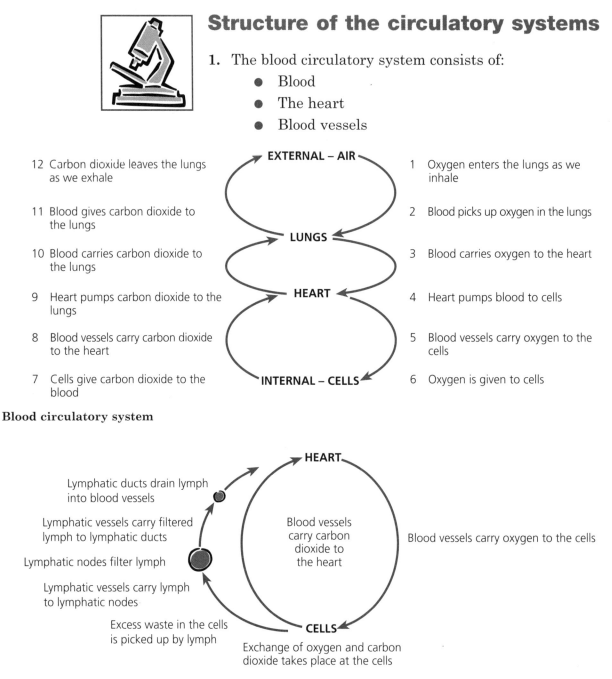

12 Carbon dioxide leaves the lungs as we exhale

11 Blood gives carbon dioxide to the lungs

10 Blood carries carbon dioxide to the lungs

9 Heart pumps carbon dioxide to the lungs

8 Blood vessels carry carbon dioxide to the heart

7 Cells give carbon dioxide to the blood

EXTERNAL – AIR

LUNGS

HEART

INTERNAL – CELLS

1 Oxygen enters the lungs as we inhale

2 Blood picks up oxygen in the lungs

3 Blood carries oxygen to the heart

4 Heart pumps blood to cells

5 Blood vessels carry oxygen to the cells

6 Oxygen is given to cells

Blood circulatory system

HEART

Lymphatic ducts drain lymph into blood vessels

Lymphatic vessels carry filtered lymph to lymphatic ducts

Lymphatic nodes filter lymph

Lymphatic vessels carry lymph to lymphatic nodes

Excess waste in the cells is picked up by lymph

Blood vessels carry carbon dioxide to the heart

Blood vessels carry oxygen to the cells

CELLS

Exchange of oxygen and carbon dioxide takes place at the cells

Lymphatic circulatory system

2. The lymphatic circulatory system consists of:
   - Lymph
   - Lymphatic vessels
   - Lymph nodes and ducts

Structures of the blood circulatory system

Blood

Blood is a fluid connective tissue containing cells that are suspended in a liquid called plasma. Blood provides a transportation medium between the internal environment of the body and the external environment of the world!

Blood consists of two parts – **plasma** and **cells**:

1. Plasma is a straw-coloured fluid which makes up approximately 55 per cent of the blood. Plasma is made up of approximately 10 per cent of proteins including: albumin, globulin, fibrinogen and prothrombin, together with 90 per cent of water in which chemical substances are dissolved or suspended e.g. waste products, nutrients, hormones, oxygen, mineral salts, enzymes, antibodies and antitoxins.

2. Cells make up the remaining 45 per cent of blood and are produced in the red bone marrow found in spongy (cancellous) bone.

There are three main types of blood cells:

1. **Erythrocytes** or red blood cells are bi concave, flexible discs. They are formed in red bone marrow and do not contain a nucleus as this disappears as the cells develop. Erythrocytes live for about 120 days, are removed from the body by the liver or the spleen and are constantly replaced by new cells. Millions of new blood cells replace these old cells everyday! Erythrocytes contain **haemoglobin** (haemo = iron, globin = protein) which contributes to the red colour of the blood. Haemoglobin is a substance that enables the erythrocytes to carry oxygen and carbon dioxide.

2. **Leucocytes** or white blood cells are colourless, irregular in shape and contain a nucleus. They are larger than red blood cells but there are fewer of them. They have a life span of between just a few hours and a few years depending on their activity. There are two types of leucocytes:

Fascinating Fact

An average adult has approximately five litres of blood in their body making up 6–8 per cent of the total body weight.

Tip

Chapter 3 provides detailed information on cancellous bone.

Fascinating Fact

The formation of **pus** at the site of an injury results from an accumulation of leucocytes which have engulfed the dirt, bacteria and dead cells before dying themselves and mixing with fluid.

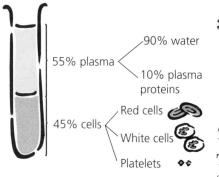

55% plasma
90% water
10% plasma proteins

45% cells
Red cells
White cells
Platelets

The blood

- *Granulocytes* – form 75 per cent of white blood cells and defend the system against viruses and bacteria. They are able to change shape and pass out of the blood vessels and into the surrounding tissue.

- *Non-granular leucocytes* (lymphocytes and monocytes). Lymphocytes are produced by lymph nodes as part of the lymphatic system and are responsible for making **antibodies** which play a major role in the body's resistance to infection. Monocytes are able to 'engulf' harmful bacteria by a process known as **phagocytosis** and successfully remove danger from circulating around the body!

3. **Thrombocytes** or platelets are much smaller than red blood cells. They are fragile, do not contain a nucleus and are associated with the formation of blood clots at the site of an injury. They are made in the red bone marrow and have a life expectancy of 5–9 days.

The heart

The heart is located in the thorax between the lungs and slightly towards the left side of the body; it is the approximate size of the owner's fist.

The heart is a hollow, muscular organ that acts like a pump. It is the centre of the blood circulatory system and is involved with the transportation of blood to and from all parts of the body via the **systemic** and **pulmonary** circulations:

- Systemic circulation involves the circulation of blood between the heart and the body via blood vessels.

- Pulmonary circulation involves the circulation of blood between the heart and lungs via pulmonary blood vessels.

The heart is made up of three layers of tissue:

- **Endocardium** or inner layer forming the lining of the heart.

- **Myocardium** or middle layer which is made up of cardiac muscular tissue. This is under **involuntary** control producing contractions in the form of the heartbeat.

- **Pericardium** or outer layer forming a double layer. The space between the layers is filled with a fluid which prevents friction and allows the layers to move freely as the heart beats.

Fascinating Fact

The heart beats approximately 130 times per minute in a new born baby reducing with age to approximately 70 beats per minute in a resting adult.

The heart is divided up into four sections or **chambers:**

- The left and right **atria** making up the upper chambers of the heart.
- The left and right **ventricles** making up the lower chambers of the heart.

A muscular wall called the **septum** separates the left and right chambers to prevent blood from the right and left sides of the body from mixing. The right side of the heart deals with blood low in oxygen known as **deoxygenated** blood, while the left side of the heart deals with blood that has a high oxygen content known as **oxygenated** blood. The atria connect with the ventricles via valves:

- The **tricuspid** valve connects the right atrium with the right ventricle.
- The **bicuspid** valve connects the left ventricle with the left atrium.

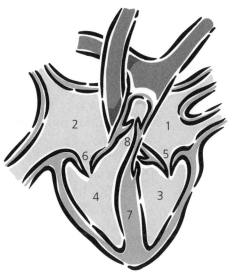

1 Left atrium
2 Right atrium
3 Left ventricle
4 Right ventricle
5 Bicuspid valve
6 Tricuspid valve
7 Septum
8 Valves

The heart

Blood vessels

Blood is circulated around the body by a network of vessels called **arteries** and **veins**. Arteries always carry blood *away* from the heart and veins always carry blood *towards* the heart.

Capillaries form at the ends of arteries and veins and provide the link between the circulatory system and the cells of the body.

- Arteries are thick-walled hollow tubes and are made up of three layers of cells. They have a fibrous outer

Tip

The blood in the arteries is known as **arterial** blood and the blood in the veins is known as **venous** blood.

Tip

Massage is normally of great benefit to venous blood flow and for this reason, movements should always be applied with the pressure aimed towards the heart. However, massage is contra indicated in the case of varicose veins. Massaging over localised swelling may cause further pain and problems especially in the case of a clot being present.

Fascinating Fact

A bruise is the result of burst capillaries at the site of an injury. Blood cells leak out into the surrounding tissue giving the skin the distinct purple colour. The colour of the bruise changes as the blood cells break down and die. The dead cells are engulfed by white blood cells before being removed from the system.

layer, a middle layer of smooth muscle and elastic tissue and an inner layer of scaly epithelial tissue. Arteries are largest as they leave the heart getting gradually smaller as they travel away from the heart. There is more elastic tissue present in the middle layer of larger arteries than smaller arteries, which tend to have more muscular tissue. This is due to the fact that larger arteries carry more blood and the elastic tissue allows the expansion of these arteries to occur. The muscular tissue helps to maintain the pressure of the blood flowing from the heart, helping to keep it moving around the body. The central cavity of arteries is known as the **lumen** and it is the walls of this that can get 'furred' up thereby preventing the free flow of blood. Arteries sub divide to form **arterioles** which are similar in structure to arteries except that they contain more muscular tissue than elastic and can relax (dilate) or contract (constrict) depending on the activity in the body. When an organ like the stomach needs more blood to start off the digestive process, arterioles dilate to allow more blood to flow to the area. When digestion stops, the arterioles constrict diverting the blood elsewhere.

- Veins are thin-walled tubes also made up of three layers, but they are thinner than arteries and do not contain as much smooth muscle or elastic tissue. Veins rely largely on the movement of the voluntary skeletal muscles to aid the flow of blood back to the heart. The inner lumen is larger in veins than in arteries. In the same way that arteries subdivide to become arterioles, veins subdivide and form venules. In addition, veins contain valves to prevent the back flow of blood as it travels towards the heart. Faulty valves result in poor venous blood circulation back to the heart and can develop into *varicose veins*. This is common in the legs where blood collects in the veins causing them to become distended and painful. Occasionally the blood may form a clot or *thrombus* which may travel through the circulatory system and cause a blockage and great harm.

- **Capillaries** are single cell structures which form a network in the tissues allowing the exchange of oxygen, carbon dioxide and nutrients to take place. The capillary walls are thin and permeable allowing substances to pass in and out. Capillaries form the end of the route for blood flow from the heart where they allow the oxygen and nutrients to pass out into the cells, and the beginning of the route of blood flow from the cells where carbon dioxide is picked up before returning back to the heart.

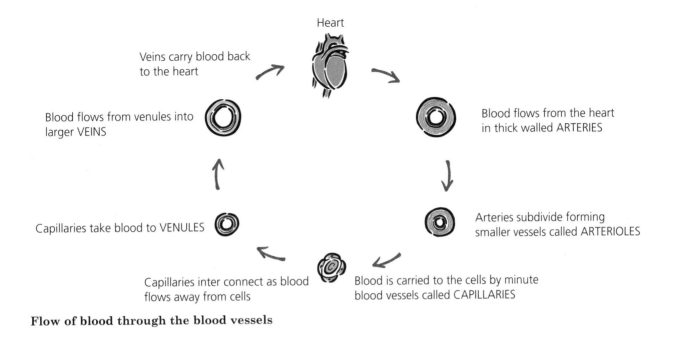

Heart

Veins carry blood back
to the heart

Blood flows from venules into
larger VEINS

Blood flows from the heart
in thick walled ARTERIES

Arteries subdivide forming
smaller vessels called ARTERIOLES

Capillaries take blood to VENULES

Capillaries inter connect as blood
flows away from cells

Blood is carried to the cells by minute
blood vessels called CAPILLARIES

Flow of blood through the blood vessels

Structures of the lymphatic circulatory system

Lymph

Lymph is a straw coloured liquid similar to blood plasma which forms as a result of the blood passing substances into the fluid which bathes the cells. This fluid is called tissue fluid or **interstitial fluid** and is formed from the blood plasma. It acts as a link between the blood and the cells allowing substances like oxygen and nutrients to pass into the cells from the blood and waste products such as carbon dioxide to pass back out. However, some of the plasma proteins also leak out into the tissues and these need to be collected back up to prevent swelling (**oedema**) from occurring in the tissues. About 10 per cent of the interstitial fluid flows into lymphatic capillaries as the walls of the lymphatic capillaries allow the plasma proteins, waste products, bacteria and viruses to pass easily through. The rest of the substances that pass out of the cells are picked up by blood capillaries and follow the course of venules and veins back to the heart.

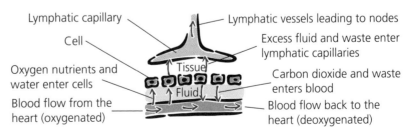

Lymphatic capillary

Cell

Oxygen nutrients and
water enter cells

Blood flow from the
heart (oxygenated)

Tissue
Fluid

Lymphatic vessels leading to nodes

Excess fluid and waste enter
lymphatic capillaries

Carbon dioxide and waste
enters blood

Blood flow back to the
heart (deoxygenated)

Lymph flow

Lymphatic vessels

Lymphatic vessels start as lymphatic capillaries which pick up the excess interstitial fluid at the tissue. They develop into larger tubes and follow the course of veins through the body. Lymphatic vessels are similar to veins as they also contain valves which prevent back-flow of lymph as it travels through the system. The flow of lymph is also assisted by the movement of skeletal muscles like the flow of venous blood.

Lymphatic nodes, tissue and ducts

Lymphatic vessels pass through nodes, tissue and eventually ducts before connecting with veins and draining into the heart for the whole process to begin again.

Lymphatic nodes

Also known as glands, they are situated in strategic places around the body. They are made up of fibrous tissue which contains different types of cells derived from white blood cells:

1. **Macrophages** – cells that destroy unwanted and harmful substances (**antigens**) by filtering lymph as it passes through the nodes.

2. **Lymphocytes** – cells that produce a defence (**antibody**) in response to an antigen picked up by the macrophages.

Lymph enters the nodes via **afferent** vessels, passes through the nodes and exits via **efferent** vessels.

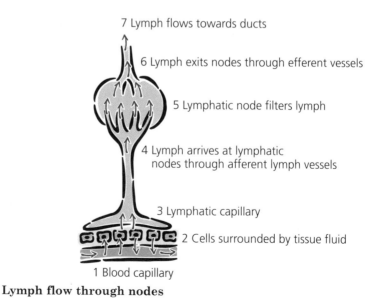

7 Lymph flows towards ducts

6 Lymph exits nodes through efferent vessels

5 Lymphatic node filters lymph

4 Lymph arrives at lymphatic nodes through afferent lymph vessels

3 Lymphatic capillary

2 Cells surrounded by tissue fluid

1 Blood capillary

Lymph flow through nodes

Tip

Chapter 5 provides more information on the tonsils and adenoids.

Fascinating Fact

If lymphatic tissue becomes chronically infected and unable to cope, it can be surgically removed and the body will adapt accordingly by using other lymphatic tissue to do the job of the lost tissue.

!

Remember

Massage can help the flow of lymph around the body in much the same way as it can help blood flow. Movements should be directed with pressure towards lymph nodes, reduced pressure over nodes and increased pressure again towards the heart. Draining movements will aid venous blood flow back to the heart as well as lymphatic flow through the system. This will have the effect of increasing the removal of waste out of the system, which is of particular importance in the case of fatigued and overworked muscles as the removal of lactic acid is speeded up helping to remove the pain from aching muscles (Chapter 4).

Lymphatic tissue

In addition to the nodes, there are other areas of the body that contain lymphatic tissue. These include:

- The **spleen** – situated in the upper left hand side of the abdomen and made up of lymphatic tissue. Its structure is similar to that of a lymph node but is larger and is the organ involved in the production of new cells and the destruction of old cells. Macrophage cells in the spleen remove old, damaged and defective erythrocytes from the blood breaking them down for recycling. The spleen also acts as a reservoir for blood. The blood stored in the spleen can be diverted to other parts of the body when needed e.g. in the case of excess bleeding (haemorrhage).

- The **thymus gland** – situated behind the sternum in the thorax. It is a site for mature lymphocytes called T-lymphocytes which are able to respond to antigens that invade the body. These T-lymphocytes are stored in the thymus and are passed into the blood stream as required.

- **Tonsils** and **adenoids** – situated in the throat (tonsils) and at the back of the nose (adenoids). They are associated with the respiratory system and help to destroy harmful invaders as we take air into the body.

- **Appendix** – situated at the start of the large intestine protecting the lower digestive system from harmful substances.

- **Lacteals** – situated in the small intestine and associated with the absorption of fat in the digestive system (Chapter 7). **Peyer's patches** in the latter part of the small intestine help to prevent infection.

Lymphatic ducts

These collect the filtered lymph that has left the nodes and drain it into the veins. There are two lymphatic ducts:

1. The **thoracic duct** – the main duct extending from the lumbar vertebrae to the base of the neck. It is approximately 40 cm long and collects lymph from the left side of the head, neck and thorax, the left arm, both legs as well as the abdominal and pelvic areas, draining it into the left subclavian vein.

2. The **right lymphatic duct** – is only 1 cm long and is situated at the base of the neck. It collects lymph from

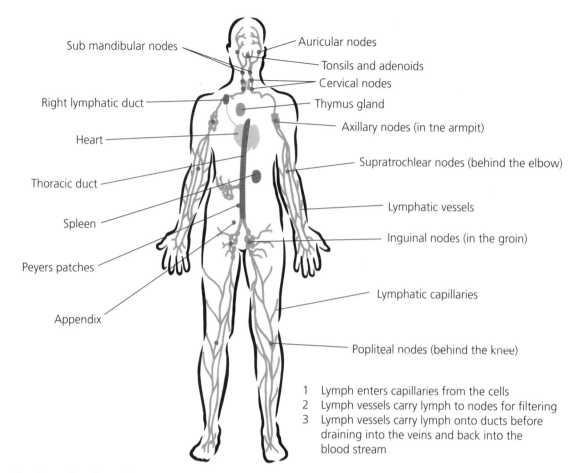

1 Lymph enters capillaries from the cells
2 Lymph vessels carry lymph to nodes for filtering
3 Lymph vessels carry lymph onto ducts before draining into the veins and back into the blood stream

Lymphatic nodes, lymphatic tissue and lymphatic ducts

the right side of the head, neck and thorax, as well as the right arm, draining it into the right subclavian vein.

Lymph then becomes part of the blood circulation once again and the whole process starts afresh.

Functions of the circulatory systems

Every living cell is reliant on the circulatory system as a whole in order to maintain its own particular functions effectively and efficiently.

There are four main functions of the system: circulation, transportation, defence and regulation.

Circulation

Circulation of blood away from the heart and to the cells of the body is controlled by the beating of the heart – each time the chambers of the heart contract and relax this may be *felt* and *heard* as a heartbeat.

 **Activity**

Think about the times your heart beats faster and you will realise that this happens whenever we need to be more physically or mentally active e.g. running for the bus, preparing for an assessment or an exam etc.

Fascinating Fact

An artificial structure called a **pacemaker** can be implanted to help the beating of the heart in cases where the heartbeat becomes irregular.

- The atria relax and fill with blood from the veins and the heartbeat can be heard as a 'lubb' sound as the valves close on the blood passing from the atria to the ventricles.
- The ventricles contract forcing the blood into the arteries and the heartbeat can be heard as a 'dupp' sound as valves close to prevent back flow.
- The relaxation is called **diastole** and the contraction is called **systole.**
- The heart beats faster when the body needs more oxygen.

The autonomic nervous system (Chapter 9) is responsible for controlling the heartbeat. By picking up on the needs of the body via nerves it is able to alert the heart and lungs. Breathing (Chapter 5) becomes faster and the rate at which the heart pumps out the incoming oxygen speeds up.

Blood pressure refers to the pumping action of the heart caused by the relaxation and contraction of the ventricles and can be measured with a **sphygmomanometer.**

- The maximum pressure associated with the contraction of the ventricles = **systolic pressure**.
- The minimum pressure associated with the relaxation of the ventricles = **diastolic pressure**.
- Normal blood pressure in an adult is approximately 120 systolic over 80 diastolic and is maintained by the force of the heartbeat, the volume of blood in the system and the resistance of blood flow in the arteries.
- High blood pressure (hypertension) occurs when the heart has to work harder to force the blood out of the left ventricle and into the **aorta** – the main artery. This puts additional strain on the heart with the potential for blood vessels to the brain to rupture causing a stroke. The actual causes of high blood pressure are linked to stress, diet, alcohol and smoking; it can also be brought on as a result of kidney disease and hardening or narrowing of the arteries; in some cases it can be hereditary.
- Low blood pressure (hypotension) occurs when the heart is unable to exert enough force on the blood as it leaves the heart resulting in insufficient blood to the brain and causing the person to feel faint. The causes of low blood pressure include hormonal and hereditary factors and can also be associated with shock.

Fascinating Fact

The resistance of the blood flow in the arteries refers to the elasticity of the arteries and the viscosity of the blood. If the arteries are damaged (become hardened and/or thickened) then blood flow may be impeded and the blood pressure will increase. If the blood is **viscous** it is thick and sticky and this also impedes blood flow and raises blood pressure. The amount of plasma proteins and erythrocytes affects the viscosity of blood.

Fascinating Fact

As soon as a person faints and falls over, their head is at the same physical level as their heart and normal blood pressure resumes!

Tip

Relaxing treatments have the effect of lowering heart rate and blood pressure. For this reason care should be taken with clients at the end of a treatment. Always ensure that a client has time to sit up slowly before getting up off the couch. This will allow the heart to adjust to the movement of the body and bring the blood pressure up to the normal rate.

Activity

It is easier to feel the pulse in the neck than in the feet because as the blood flows further from the heart, the pressure is less than that which propelled the blood out of the heart and into the arteries.

The contraction and relaxation of the ventricles can be felt as a **pulse**. It is in fact the wave of *pressure* as the blood flows through the arteries, arterioles and capillaries to the cells. If an artery is pressed against a bone the pulse can be felt.

The pulse *rate* relates to the speed of the heartbeat and the *strength* of the pulse relates to the subsequent pressure of blood flow out of the heart. The pulse rate is affected in much the same way as blood pressure, increasing with activity and decreasing during rest. The normal resting pulse rate for an adult is between 70–80 beats per minute increasing to between 180–200 beats a minute during maximum activity.

Circulation of blood and lymph back to the heart is controlled by:

- The action of the skeletal muscles (Chapter 4). As the muscles contract and relax they squeeze the blood through the veins and lymph through the lymphatic vessels.
- The position of valves in veins and lymphatic vessels prevent backflow.

Circulation of blood and lymph through the body is continuous but may be divided into two main parts – **pulmonary** and **systemic** circulation with **portal** (relating to the digestive system) and **coronary** (relating to the heart) circulation being subsidiaries of the systemic circulation.

Pulmonary circulation involves the flow of blood between the lungs and the heart:

- Four pulmonary veins (two from each lung) carry freshly oxygenated blood to the left atrium of the heart. It then passes through the bicuspid valve into the left ventricle ready to be pumped around the body.
- The left and right pulmonary arteries carry deoxygenated blood from the right ventricle to the lungs where the carbon dioxide is expired and replaced with oxygen.

Systemic circulation involves the general flow of blood from the heart and the return of blood and lymph from the cells.

- Oxygenated blood passes from the left atrium through the bicuspid valve into the left ventricle and is pumped out of the heart to the cells of the body via the aorta (main artery). From there the blood flows to the head via the carotid arteries, the arms via the subclavian, axillary, brachial, radial and ulna

Blood vessels

A   Main artery AORTA taking oxygenated blood *away* from the heart to the cells

B   Pulmonary artery taking deoxygenated blood *away* from heart to lungs

C   Superior and inferior vena cava (veins) bringing deoxygenated blood *to* the heart

D   Pulmonary veins bringing oxygenated blood *to* the heart

The heart

1   Left atrium receives oxygenated blood from the lungs

2   Right atrium receives deoxygenated blood from the cells

3   Left ventricle sends blood to the cells

4   Right ventricle sends blood to the lungs

5   Bicuspid valve separates upper and lower chambers preventing back flow of blood

6   Tricuspid valve separates upper and lower chambers preventing back flow of blood

7   Septum separating left and right sides of the heart

The heart and main blood vessels

arteries, and the legs via the iliac, femoral, popliteal and anterior tibial arteries.

● Deoxygenated blood is taken to the right atrium of the heart by the main veins of the body. These include the anterior tibial, popliteal, femoral and iliac veins from the legs, the radial, ulna, brachial axillary and subclavian veins from the arms and the jugular veins from the head. These all drain blood into the superior and inferior vena cava (main veins) where it passes from the atrium, through the tricuspid valve into the right ventricle.

● Lymph flows in lymph vessels alongside the deoxygenated blood flowing in veins and is filtered through nodes – popliteal behind the knees, inguinal in the groin, supratrochlear behind the elbow, axillary in the armpits and cervical, submandibular, auricular and occipital in the head and neck, before being collected by the right lymphatic and thoracic ducts and drained into the subclavian veins prior to its entry into the heart.

● Portal circulation involves the flow of blood from the digestive system to the liver via the portal vein in order to modify and regulate the supply of nutrients to all the parts of the body.

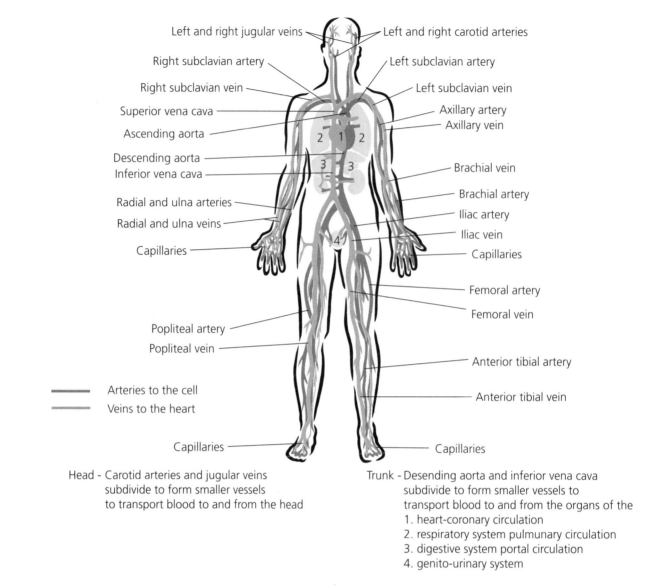

Left and right jugular veins

Left and right carotid arteries

Right subclavian artery

Left subclavian artery

Right subclavian vein

Left subclavian vein

Superior vena cava

Axillary artery

Axillary vein

Ascending aorta

Descending aorta

Brachial vein

Inferior vena cava

Brachial artery

Radial and ulna arteries

Iliac artery

Radial and ulna veins

Iliac vein

Capillaries

Capillaries

Popliteal artery

Femoral artery

Popliteal vein

Femoral vein

Anterior tibial artery

Anterior tibial vein

Arteries to the cell

Veins to the heart

Capillaries

Capillaries

Head - Carotid arteries and jugular veins
subdivide to form smaller vessels
to transport blood to and from the head

Trunk - Desending aorta and inferior vena cava
subdivide to form smaller vessels to
transport blood to and from the organs of the
1. heart-coronary circulation
2. respiratory system pulmunary circulation
3. digestive system portal circulation
4. genito-urinary system

The main arteries and veins of the body

- Coronary circulation involving the flow of blood to and from the heart itself via coronary arteries and veins to ensure that it has an adequate supply of nutrients to carry out the vital functions demanded of it.

Changes in blood volume in different areas of the body results in **blood shunting.** Blood is directed to those areas of the body that need it according to the physical demands of the particular system or organ e.g. the volume of blood is greater at the digestive system just after a meal than at the muscles in order to aid the digestive process. Treatments are contra indicated just after a heavy meal, as blood would be directed away from the digestive system to the muscles being worked on, causing digestive problems.

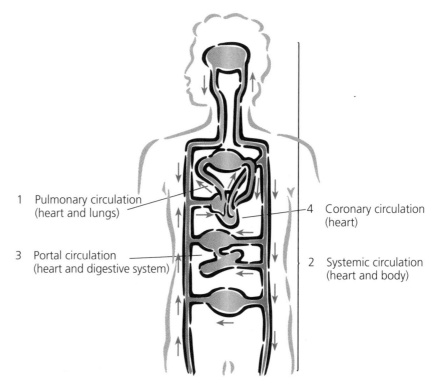

1   Pulmonary circulation
    (heart and lungs)

4   Coronary circulation
    (heart)

3   Portal circulation
    (heart and digestive system)

2   Systemic circulation
    (heart and body)

Circulation

Transportation

Substances are transported around the body by the blood.

- Erythrocytes transport oxygen and carbon dioxide between the lungs and the cells of the body with the aid of **haemoglobin.** The haemoglobin combines with oxygen as we inhale becoming *oxy*haemoglobin. It is bright red in colour and takes oxygen to the cells suspended in the blood via arteries. Carbon dioxide replaces the oxygen in the haemoglobin making it *deoxy*haemoglobin. It then becomes dark red in colour as the blood returns to the lungs via veins and we exhale the carbon dioxide out of the body.

In addition to the transportation of oxygen and carbon dioxide, other substances suspended or dissolved in blood plasma are transported around the body:

- Waste products from the cells, e.g. **urea** are transported to the organs of excretion i.e. the liver, kidneys and sweat glands and eliminated out of the body in sweat or urine.

- Hormones released from endocrine glands act as chemical messengers instructing different parts of the body to perform vital functions. The blood transports them to the relevant body systems when

Tip

It is therefore not advisable to run and eat at the same time as the muscles would require a greater volume of blood and the result is likely to be muscles that tire easily together with indigestion!

Remember

Veins appear blue because of the deoxygenated blood they carry.

needed e.g. the hormone **adrenalin** is transported from the adrenal glands to activate the muscles when we need to flee from a dangerous situation.

- Nutrients and water from the digestive system are transported to the cells for cellular metabolism. This process 'feeds' the cells enabling them to carry out their functions of reproduction and repair.

- Minerals from the digestive system and those produced within the body are transported to the cells of the body to help maintain their acid pH balance as well as their individual functions. Minerals include sodium chloride, sodium carbonate, potassium, magnesium, phosphorus, calcium, iodine and copper.

- Enzymes, which are proteins, are produced in cells and have the ability to produce or speed up chemical changes without changing themselves. These chemical catalysts are transported in the blood e.g. the pancreas produces enzymes which are transported by the blood to the small intestine where they aid in the process of digestion.

- Antibodies and antitoxins are transported from the lymph nodes where they are produced in response to an invasion within the body from the toxins released by bacteria or viruses. The blood transports them to the site of invasion.

Lymph transports:

- Waste products and tissue fluid from the cells to lymph nodes for filtering
- Fluid from the lymph nodes to the ducts to be drained back into the blood
- Fats from the digestive system to the blood stream.

> ### Tip
>
> Increasing blood and lymph circulation will speed up the transportation of these substances allowing the body to perform more efficiently and effectively.

Defence

The circulatory system plays an important role in the defence of the body:

- Leucocytes (white blood cells) help to destroy damaged and old cells. To help defend the system against viruses and bacteria, some of the leucocytes are able to increase in number by mitosis in order to over run the invasion!

- Lymph nodes provide a filtering system for lymph as macrophages and lymphocytes are able to engulf antigens and produce anti bodies in defence.

Tip

Stimulation of blood flow through massage can also increase the flow of blood to an area with the compression action on the blood vessels. This causes the blood to rush to the surface of the skin and reddening of the skin to occur (erythema). This stimulation in circulation ensures that the surface structures receive an increased supply of oxygen and nutrients aiding skin (Chapter 2) and muscles (Chapter 4).

- The spleen filters blood in much the same way as the nodes filter lymph, adding to the defence mechanisms of the system as a whole.

- Blood clotting occurs to seal the site of an injury to prevent excess blood/fluid loss. Thrombocytes (platelets) perform this vital function by releasing an enzyme which changes the plasma proteins to form a protective structure over the damaged area. This acts as a trap for white blood cells and platelets which form a clot at the site of the injury. The blood clot dries to form a scab which gives added protection to the area until healing of the tissues has taken place. The scab is then replaced with new cells.

- In the case of an allergic reaction or if the skin is damaged, the blood flow to the area will increase so that the blood cells can deal with the problem. This reddening of the skin is known as **erythema**.

Regulation

The circulatory system contributes to the overall homeostasis of the body in the following ways:

- Body processes are regulated by the hormones carried in blood.

- The pH of blood is maintained between 7.35–7.45 by chemical systems in the blood called buffers. A significant rise (Alkalosis) or fall (Acidosis) in blood pH can be fatal.

- Normal fluid balance is maintained by the composition of the blood.

- The correct body temperature of 36.8°C is maintained by the blood through the transportation of heat. The heat comes from the activity in the muscles and organs such as the liver. The blood is able to distribute this heat to different areas of the body by vasodilation and vasoconstriction of blood vessels.

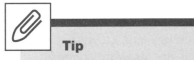

Tip

Chapter 2 provides more detail on vasodilation and vasoconstriction.

The circulatory systems provide the force that links the systems of the body while blood contains the necessary components for life.

Common conditions

An A–Z of common conditions affecting the circulatory systems

- ACROCYANOSIS – deficiency in the circulation of the hands and feet.

- AIDS – Acquired Immune Deficiency Syndrome caused by the HIV Human Immune Deficiency virus. The T-lymphocytes are attacked making the immune system incapable of functioning effectively.

- ANAEMIA – a decrease in the production of red blood cells and/or haemoglobin. This may be caused by excessive blood loss reducing the amount of red blood cells in the body, lack of iron affecting the haemoglobin function or dysfunction in the bone marrow resulting in loss of production of new blood cells.

- ANEURYSM – localised swelling of an artery which can develop if the artery is diseased or weakened, especially if blood pressure is high.

- ANGINA – reduction of blood flow to the heart usually brought on by excessive exertion.

- ARTERIAL THROMBOSIS – clotting of blood in an artery obstructing normal blood flow.

- ARTERIOSCLEROSIS – the walls of the arteries lose their elasticity and harden. This has the effect of raising the blood pressure.

- ARTERITIS – inflammation of an artery often associated with rheumatoid arthritis.

- ATHEROSCLEROSIS – a narrowing of the arteries caused by a build up of fats including cholesterol.

- ATRIAL FIBRILLATION – irregular heart beat.

- BLUE BABY – heart disorder present at birth resulting in some of the blood not passing through the lungs to receive oxygen.

- CORONARY THROMBOSIS – a common cause of heart attacks where an artery supplying the heart is obstructed.

- DIABETES – a condition whereby the body cannot use the sugars and carbohydrates from the diet. The hormone insulin, which is produced by the pancreas, helps to regulate the use of sugars and carbohydrates by the body. Diabetes occurs when the pancreas fails to produce enough insulin resulting in a build-up of sugar in the blood.

- GANGRENE – lack of blood supply to the fingers and toes results in decay and eventual death of the part.

- HAEMOPHILIA – the blood is not able to clot resulting in excessive blood loss.
- HAEMORRHOIDS – varicose veins in the rectum or anus. More commonly referred to as piles.
- HEART FAILURE – the heart weakens as a result of not being able to cope with the pumping action.
- HEPATITIS B or C – inflammation of the liver caused by different viruses that are transmitted by infected blood.
- HIGH CHOLESTEROL – causes a build up of the fatty substance cholesterol in the arteries which results in ATHEROSCLEROSIS and HIGH BLOOD PRESSURE.
- HODGKIN'S DISEASE – cancer of the lymphatic tissue.
- HYPERTENSION – high blood pressure.
- HYPOTENSION – low blood pressure.
- LEUKAEMIA – over-production of white blood cells resulting in cancer of the blood.
- LYMPHOEDEMA – swelling in a limb which causes the lymphatic circulation to become affected.
- OEDEMA – swelling caused by excess fluid from the circulatory system which has accumulated within the tissues.
- PHLEBITIS – inflammation of a vein commonly affecting the legs.
- PULMONARY EMBOLISM – blockage of the blood vessels in the lungs.
- RAYNAUD'S SYNDROME – contraction of the arteries supplying blood to the hands and feet causing numbness.
- RHEUMATIC FEVER – inflammation of the heart which often follows tonsillitis.
- SEPTICAEMIA – blood poisoning caused by the multiplication of harmful substances in the blood.
- STRESS – causes the heart to beat faster raising the pulse rate and blood pressure. Excessive stress levels can lead to heart problems.
- THROMBO PHLEBITIS – inflammation of the length of a vein usually occurring in the legs.
- THROMBUS – a blood clot in the blood vessels or the heart.
- VARICOSE VEINS – ineffective valves in veins which cause the blood to collect in the veins instead of returning to the heart. This results in veins becoming distended and painful.

System sorter

CIRCULATORY SYSTEM

Skeletal

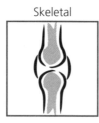

Red bone marrow in cancellous bone is responsible for the development of cells found in blood and lymph.

Muscular

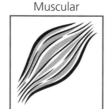

The action of the skeletal muscles helps stimulate venous blood flow to the heart and lymphatic flow towards the lymph ducts.

Integumentary

Body temperature is adjusted through blood flow to and from the skin – vasodilation and vasoconstriction.

Respiratory

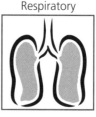

The respiratory system oxygenates and deoxygenates blood in the lungs. Lymphocytes are responsible for guarding the respiratory system against infection.

Lacteals in the small intestine absorb fats from the digested food and Peyers patches guard against infection. The blood absorbs the other nutrients and water transporting it to the cells.

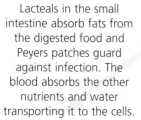

Digestive

The kidneys clear the blood of waste products and excess fluid. They also control the water levels of the blood and lymph releasing less urine when water levels are low.

Genito-Urinary

The nervous system controls blood flow and blood pressure by 'picking up' on the needs of the body and activating an increase or decrease in heart rate.

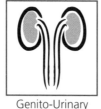

Nervous

The blood carries hormones produced by the endocrine glands around the body to activate body changes according to body needs.

Endocrine

The circulatory system consists of two complementary systems, blood circulation and lymphatic circulation. As a combined system they are responsible for

- Circulation of blood and lymph
- Transportation of substances eg oxygen and carbon dioxide
- Defence against disease
- Regulation of body processes

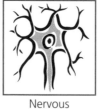

Holistic harmony

Remember

Caffeine has the effect of stimulating the heart rate in much the same way as stress. For this reason caffeine provides us with a 'quick fix' but as a result can have a long-term debilitating effect on the well-being of the body in the same way that stress does.

Remember

Salt should be avoided in excess as this adds to the increase in blood pressure.

Angel advice

Aspirin is often taken to thin the blood and reduce the risk of heart attacks. It has been found that taking a vitamin E supplement can often be equally effective.

The blood and lymphatic circulatory systems are the link between all of the systems and provide every cell with the fundamental components vital for life in the form of oxygen, nutrients and water. The circulatory systems also provide a 'pick up' and a 'messaging' service by collecting the waste products produced by the cells and passing on hormones to instruct the cells. In order to perform these vital functions efficiently and effectively the circulatory systems require a certain amount of care so that homeostasis can be maintained.

Fluid

As with all systems, the circulatory system relies on the balance of fluid in the body.

- The volume of blood in the body is affected by the amount of fluid taken into the body. If the body is lacking in fluid and is allowed to dehydrate the blood volume will also fall. This results in reduced blood pressure and fainting.
- The volume of lymph in the body is also affected by the amount of fluid taken into the body. Dehydration results in 'static lymph' which hinders the normal flow resulting in puffiness and oedema (swelling).
- Lack of fluid affects the water content of plasma making the consistency of blood more viscous. This restricts blood flow and raises blood pressure.

Nutrition

Whilst the circulatory system transports nutrients to other parts of the body, it also relies heavily on the dietary intake of a person for its own well-being. It requires the same balanced diet as other body systems with a high intake of anti oxidants, in particular vitamin C which also helps to keep arteries supple. Other useful nutrients include:

- *Iron* – needed for the development of haemoglobin in the red blood cells and found in pumpkin seeds, parsley, almonds, cashew nuts and raisins.
- *Folic acid* – needed for the development of red blood cells generally. The best food sources include wheatgerm, spinach, peanuts and sprouts.
- *Vitamin B12* – assists with the transportation of oxygen within the blood and is found in oysters, sardines and tuna fish.

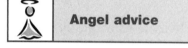

Rest

The circulatory system slows down during rest. The heart beats more slowly, the blood pressure is reduced and the pulse rate is slower and weaker as a result. Blood and lymph flow slows down as oxygen uptake is reduced. It is also worth remembering that venous blood and lymph have to defy gravity on their route back to the heart and when we are lying down the challenge is not so great! Lying down with the legs slightly raised helps even further as this actively encourages the return flow. Rest is vital to counteract the effects of activity but can be harmful to the system as a whole if taken in excess. Sedentary people have a greater risk of problems associated with the circulatory system than active people and this risk increases with age, poor diet, lack of fresh air and stress.

Activity

The maintenance of the circulatory systems relies on regular activity to aid in the flow of venous blood to the heart and of lymph to the nodes, ducts and veins. The system responds to regular, progressive exercise much better than it does to sudden bursts of activity. A 20 minute workout three times a week is recommended to raise the heart rate and stimulate the uptake of oxygen and the release of waste in the cells. Problems occur when excessive demands are put on the system suddenly. The result of unaccustomed exercise can be damaging to the heart. For exercise to be of benefit to the system, the heart rate should not exceed 85 per cent of its 'theoretical maximum'.

Jumping exercises such as trampolining, star jumps etc. are especially useful for moving blood and lymph around the body, and upper body movements that expand the chest are good for the heart and the thoracic duct. It is also important not to underestimate the benefits associated with everyday activities like walking, going up and down stairs and even housework in helping to keep the body active and the circulatory systems fit!

Air

Certain gases inhaled into the body affect the haemoglobin in the erythrocytes (red blood cells) making the transportation of oxygen difficult e.g. carbon monoxide. Cigarette smoke produces small amounts of carbon monoxide which contributes to the damaging effects of smoking. The defective haemoglobin stimulates the production of extra erythrocytes in an attempt to rectify the situation. Thus the damaging effects of one cigarette may be counteracted by this stimulation of blood cells, but the long-term effects of smoking go beyond the healing capabilities of the body as blood pressure increases, and illness may follow. Stimulation of erythrocytes also occurs when the body is exposed to high altitude. The lack of oxygen in the air results in low blood oxygen levels which stimulates the red bone marrow to produce more erythrocytes. This increase of cells containing haemoglobin increases the uptake of oxygen bringing the levels of blood oxygen back to normal. If blood oxygen levels increase again then erythrocyte production decreases thus maintaining homeostasis. It is for this reason that the body sometimes takes time to adjust to a new environment e.g. high or low altitude.

The simple act of breathing helps the flow of lymph through the vessels. The bellowing action of the lungs helps to massage the thoracic duct aiding lymph flow to the area. Deep breathing multiplies the effect as the pressure variance in the thorax stimulates the lymphatic flow further helping to release waste products more quickly. This prevents a build up of waste in the body and eliminates any associated problems e.g. oedema.

Angel advice

Ageing need not be all doom and gloom. The passing of time is usually accompanied by people having more time to devote to the care of themselves. This time and care can be further enhanced with a positive outlook on life, which will make the latter years of life more enjoyable!

Age

Age affects the circulatory system in the following ways:

- Blood pressure may rise due to poor diet, alcohol, stress etc. and this may contribute to the risk of heart problems.
- The supply of oxygen to the cells is reduced due to the fall in air being taken into the lungs, causing breathlessness to increase with age.
- Reduced supply of oxygen affects cellular respiration (chapter 5) and contributes to poor skin condition and muscle tone.
- As the body slows down so does the circulation generally, affecting the body's defence mechanisms.

Colour

Red is associated with oxygenated arterial blood and indigo with deoxygenated venous blood. Red as a colour is energising and uplifting and indigo is calming and cooling. Consequently red is said to be helpful for anaemia and low blood pressure and indigo is helpful for haemorrhaging and

Tip

Colour is all around us and we are often intuitively drawn to a certain colour or shade depending on our physical and emotional needs.

Angel advice

It becomes the responsibility of every adult to take reasonable care of his or her own body and to contribute to the care of others in order to give everyone the best possible chance of achieving a long and healthy life.

Angel advice

As therapists we should learn to 'practise what we preach' if we are to gain the respect of our clients.

high blood pressure. The colour green represents the fourth chakra with the heart and thymus gland being its focal points. The heart relates specifically to the blood circulation and the thymus to the production of lymphocytes for the lymphatic system. When we talk about our inner most feelings, we will often physically touch the area of the heart and so green has an effect on our true self. As green is the colour found in the middle of the spectrum of rainbow colours, it also represents balance. An imbalance of green in our lives, e.g. living in built-up cities with little areas of greenery, is thought to be a contributory factor in the balance of our feelings and it is for this reason that we often feel re-energised by a trip into the country or a walk in the park.

Awareness

The circulatory system relies on the overall care of the body to keep it functioning effectively. A person will thrive both mentally and physically when surrounded by care. Think about the difference a caring treatment, a thoughtful boss or a loving partner makes to our general well-being. A treatment improves skin colour and praise from your boss will make you 'flush' with pride in much the same way as a loving gesture makes you feel 'warm' inside. This is all associated with the stimulation of the circulatory systems which is vital if we are to stay healthy. In contrast, stress raises blood pressure and heart rate and can put excessive strain on the systems. It is therefore necessary to become more aware so that we are able to identify and eliminate the excessive stresses from our lives, which in turn will allow the systems to function better and for longer.

Special care

Blood is often associated with our personality. We may refer to a person's individual characteristics by talking about 'bad' and 'good' blood and strong emotions often evoke such phrases as 'the thought made my blood boil' or 'the sound makes my blood curdle'. This highlights the connection between body and mind and demonstrates how the two are intrinsically linked and form a complete network. If the much sought-after balance is to be achieved between body and mind, the needs of the circulatory system cannot be ignored. Special care is really common sense, a working knowledge of the structures and functions of the circulatory systems will provide a level of awareness that should not be overlooked if we are to make the most of the human body as a whole and encourage our clients to do the same.

Treatment tracker

CIRCULATORY SYSTEM

Make up

Green corrective make up can be used to camouflage broken capillaries on the face or body.

Facials

Static lymph contributes to puffy, problem skin. Facials can improve blood and lymph flow eliminating such problems.

Nail care

Hand and foot massage improves blood and lymph flow to and from the germinating matrix thus improving the growth and condition of the nails.

Hair removal

Blood and lymph flow is increased when hairs are first removed from the skin due to the stimulation at the matrix. Repetitive removal will eventually 'wear down' the stimulating effect of the blood/lymph flow.

Massage stimulates blood and lymph flow. Massage movements should be directed towards the lymph nodes for drainage and towards the heart for venous blood return.

Massage

The balancing effect of reflexology results in prescribed medication being utilised more efficiently by the body as the blood circulates up to the relevant systems.

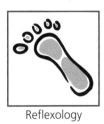

Reflexology

Geranium, juniper and rosemary are all essential oils that stimulate the lymphatic system. Black pepper and marjoram have a stimulating effect on blood flow.

Aromatherapy

Vacuum suction facials and body treatments are very effective at draining static lymph reducing puffiness as well as stimulating blood and lymph flow aiding cellular metabolism.

Electrical

Beauty and holistic treatments are of great benefit to the circulatory systems of the body helping to stimulate blood and lymph flow which in turn has a beneficial effect on every cell, tissue, organ and system.

Knowledge review – Circulatory system

1   Name the two complementary systems that make up the circulatory system.

2   What are the two main components of blood?

3   What are erythrocytes, leucocytes and thrombocytes?

4   What do erythrocytes contain and why?

5   What are the basic functions of leucocytes and thrombocytes?

6   Name the chambers of the heart.

7   Which main blood vessels carry blood away from the heart and which carry it towards the heart and what do they sub divide into?

8   What structures prevent back flow of blood in the heart, arteries and veins leaving the heart, veins and lymph in lymphatic vessels?

9   Name the smallest blood and lymphatic vessels.

10   Name four substances transported by the blood.

11   What is the coronary circulation involved with?

12   What is the main purpose of lymph?

13   What are the main functions of lymph nodes?

14   Name four other areas of the body that are comprised of lymphatic tissue.

15   Name the two lymphatic ducts.

16   What do the lymphatic ducts drain into?

17   Define the terms systolic blood pressure and diastolic blood pressure.

18   How can a pulse be felt in the arm?

19   Why does the heartbeat increase during exercise?

20   Name the four functions of the circulatory system.

The digestive system

After reading this chapter you should be able to:

- **Recognise the structures that make up the alimentary canal and the accessory organs**

- **Identify the different processes involved with the intake of food and fluid**

- **Understand the functions of the digestive system**

- **Be aware of the factors that affect the well-being of the digestive system**

- **Appreciate the ways in which the digestive system works with the other systems of the body to maintain homeostasis.**

This chapter continues to focus on the input and output functions of the body with its utilisation of nutrients and water. The system responsible for processing the food that we eat and the fluid that we drink is **THE DIGESTIVE SYSTEM**. The digestive system starts at the mouth extending down into the area of the trunk below the diaphragm; it is responsible for supplying the body with vital nutrients for growth, maintenance and general well-being. It takes the food and fluid we consume during the course of each day and transforms it into minute particles suitable for use by the cells. The digestive system is reliant on the circulatory systems (Chapter 6) for the transportation of these nutrients around the body. The digestive system is also responsible for sorting out the parts of food and fluid that are not needed by the cells, preparing it for elimination out of the body at the end of the process.

Science scene

Structure of the alimentary canal, liver, gall bladder and pancreas

The digestive system starts at the mouth and ends at the large intestine; it is known collectively as the **alimentary canal**.

The alimentary canal consists of the following:

- Mouth
- Pharynx

A Parotid salivary gland

B Sublingual salivary gland

C Submandibular salivary gland

Upper digestive system

Mouth and tongue

Nasopharynx

Oropharynx

Laryngo pharynx

Oesophagus

Middle digestive system

Liver

Stomach

Gall bladder

Pancreas

Small intestine

Small intestine

Lower digestive system

Large intestine

Small intestine

Large intestine

Anus

- Oesophagus
- Stomach
- Small intestine
- Large intestine.

In addition to the alimentary canal, the digestive system also relies on three *accessory organs*:

The accessory organs include:

- Gall bladder
- Pancreas
- Liver.

Structures of the alimentary canal

The alimentary canal can be subdivided into three sections: the upper section consisting of the mouth and pharynx, the middle section consisting of the oesophagus and stomach and the lower section consisting of the small and large intestines.

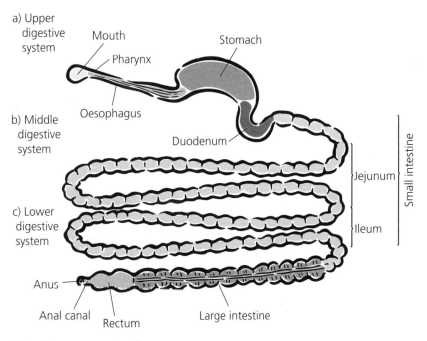

The alimentary canal

Upper digestive system

Mouth

The mouth is the first part of the digestive system and comprises the following structures: the hard and soft palate, lips, muscles, teeth, salivary glands and tongue.

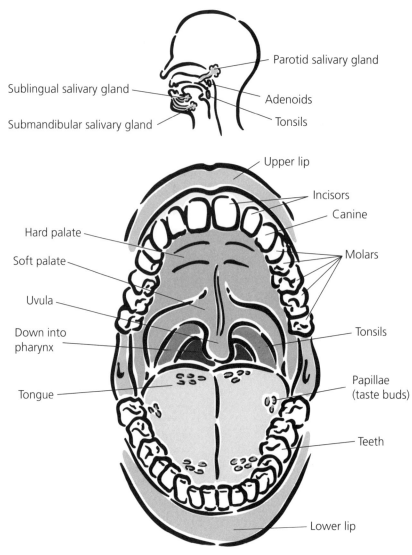

The mouth

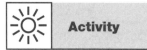

Activity

Take the tip of your tongue along the roof of your mouth and you will feel the difference between the hard and soft palate.

- Hard and soft palate – forming the roof of the mouth. The hard palate is made up of the maxilla and palatine bones (Chapter 3) and lies at the front of the mouth. The soft palate is made up of muscle and lies at the back of the mouth forming an arch called the **uvula**.

- Lips – extremely flexible structures forming the opening of the mouth. They are made up of muscle and contain a rich blood supply giving them their distinctive colour as well as a nerve supply which enables them to detect the temperature of food and fluid as it enters the mouth.

- Muscles – there are three main facial muscles responsible for **mastication** (the chewing action):
  1. Buccinator muscles of the cheeks
  2. Masseter muscles at the sides of the face

Remember

An enzyme is a protein catalyst. This means that it is made up of a protein and is responsible for making chemical changes in the body without changing itself.

Tip

Chapter 4 provides more information regarding the muscles of this area.

3. Temporalis muscles at the temple region of the face.

- Teeth – start as 20 primary teeth in small children and are gradually replaced by 32 secondary teeth between the ages of six and 25. They are divided into 16 upper teeth embedded in sockets in the upper jaw (maxillae bones) and 16 lower teeth embedded in sockets in the lower jaw (mandible bone). There are three different types of teeth:

  1. Front incisor teeth which are chisel shaped
  2. Middle canine teeth which are cone shaped
  3. Back pre molar and molar teeth which are flatter in shape.

- Salivary glands – contain cells, which produce thick watery fluid called saliva. Saliva consists of water, mucus and the enzyme salivary amylase.

There are three pairs of salivary glands:

  1. Parotid glands situated below the ear
  2. Sublingual glands situated below the tongue
  3. Submandibular glands situated below the mandible bone.

- Tongue – is attached to the hyoid and mandible bones and is made up of skeletal muscle. Its surface consists of tiny projections called **papillae** which contain cells that are sensitive to taste. For this reason they are also known as **taste buds**.

Pharynx

The pharynx has links with the respiratory and digestive systems through its three main parts:

1. *Nasopharynx* – providing the passageway for air breathed in through the nose. It is associated with the respiratory system (Chapter 5) rather than the digestive system.

2. *Oropharynx* – opens below the soft palate and nasopharynx providing a passageway for air, food and fluid that has been taken in through the mouth.

3. *Laryngopharynx* – carries on from the oropharynx providing a passageway down into the alimentary canal.

Tonsils in the throat and adenoids at the back of the nose protect the body against infection entering the body via food, fluid and air.

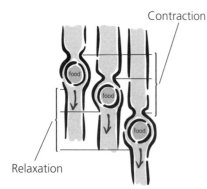

Contraction

Relaxation

Peristalsis

Middle digestive system and lower digestive system

The middle and lower sections of the alimentary canal contain a general structure along its length from the oesophagus through to the anus. It is modified at certain points depending on its specific functions.

The general structure comprises of four main layers:

1. **Peritoneum** – tough outer layer which secretes a lubricating fluid allowing the organs of the digestive system to slide against each other.

2. *Layers of muscle* – muscle fibres are arranged in two layers. The innermost layer is comprised of circular muscles and the outermost layer is comprised of long muscles arranged longitudinally. The action of these muscles as they contract and relax is known as **peristalsis** and refers to the wave of movement as ingested food travels along the alimentary canal.

3. **Submucosa layer** – consists of loose connective tissue with some elastic fibres containing blood, lymph vessels and nerves which contribute to the general welfare of the alimentary canal through 'feeding' and 'feeling'.

4. **Mucosa layer** – the innermost layer that produces a constant supply of mucus to protect the walls of the alimentary canal. This layer also produces various digestive 'juices' which contribute to the breaking down of food as it passes through the canal.

Modifications of this basic structure occur in the following parts of the alimentary canal:

Oesophagus

The oesophagus is a long tube (measuring approximately 25 cms) which extends from the pharynx to the stomach. It lies behind the trachea (windpipe) and in front of the spine. When empty, the oesophagus is a flattened tube. Its muscular structure allows it to extend as swallowed food enters. The muscular layer contracts to move the food down the oesophagus in a peristaltic action through a ring of muscle called the **cardiac sphincter** and into the stomach.

Stomach

The stomach is a J-shaped sac which lies under the diaphragm on the left side of the body. The mucosa layer of the stomach has lots of folds called **rugae** that allow it to

Fascinating Fact

A peristaltic rush occurs when the alimentary canal becomes irritated. A wave of movement forces the contents of the lower part of the alimentary canal through at great speed resulting in diarrhoea. Absorption of nutrients and water fails to take place in the 'rush' and the body very quickly becomes dehydrated and weak.

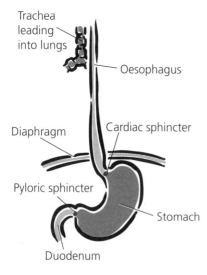

Trachea leading into lungs

Oesophagus

Diaphragm

Cardiac sphincter

Pyloric sphincter

Stomach

Duodenum

Middle digestive system

Fascinating Fact

If you think about the structure of the trachea you will remember that it is made up of cartilage at the front and soft tissue at the back. It is this soft tissue that allows the oesophagus to change shape easily as the food passes through.

Activity

Vomiting happens as a result of irritation in the stomach which sets off a chain reaction forcing the food out of the stomach through the cardiac sphincter, back up the oesophagus, past the pharynx and out of the mouth! An unpleasant feeling, but a very clever way of protecting the body against further irritation!

stretch out when full and contract when empty. Gastric glands are also present in this layer and are responsible for producing gastric juices to liquefy the food.

The muscular layer of the alimentary canal is thickest in the stomach and is responsible for a churning action which helps the liquefying process. At the end of the stomach is another ring of muscle called the **pyloric sphincter**. This muscle controls entry of liquefied food into the lower parts of the digestive system.

Small intestine

The small intestine is not small at all. In fact it is very long – approximately 6 metres. It coils around itself and fills the bulk of the abdominal cavity. It is made up of three sections: the **duodenum**, the **jejunum** and the **ileum**.

The basic structure of the small intestine is the same as the rest of the alimentary canal except that the inner mucosa layers consist of tiny projections called **villi**. Situated within the villi are glands which secrete intestinal juices to help the digestion of food; blood capillaries which absorb nutrients from the digested food; and lymphatic capillaries called **lacteals** which absorb fats from the digested food. Solitary lymph nodes are also present throughout the small

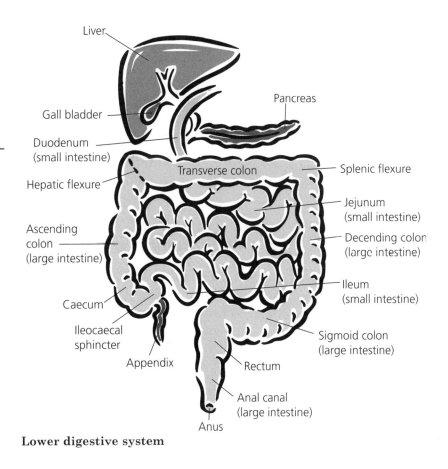

Lower digestive system

intestine and larger nodes known as **Peyer's patches**, situated in the latter part of the ileum, help to guard the system against infection (Chapter 6).

The small intestine is also linked to the accessory organs. The gall bladder and pancreas both join the small intestine at the duodenum via the bile and pancreatic ducts respectively. These organs help the process of digestion and absorption of nutrients to take place.

At the very end of the ileum there is a ring of muscle called the **ileocaecal sphincter** which separates the small and large intestines.

Large intestine

The large intestine is wider and shorter than the small intestine. Also known as the bowel or colon, it is about one and half metres long and is divided into five sections which form a rectangle around the small intestine – the caecum, the colon, the rectum, the anal canal and the anus.

- The caecum is separated from the ileum of the small intestine by the ileocaecal sphincter. Attached to the caecum is the appendix, which is made up of lymphatic tissue. It has no known digestive function but contributes to the protection of the system.
- The colon is divided into four parts – ascending, transverse and descending, which basically describe its position, ending with the sigmoid colon which links with the rectum.
- The rectum leads on from the sigmoid colon and lies next to the sacral bones of the spine.
- The anal canal leads on from the rectum.
- Finally, the anus forms the end of the large intestine providing an opening out of the body. It consists of two muscles – the internal and external sphincter muscles.

Structures of the accessory organs

The liver, gall bladder and pancreas are all organs which have an effect on the digestive system. They also have important functions associated with other systems, which makes each of them a vital link in the smooth running of the body as a whole.

Liver

The liver is the largest internal organ of the body. It lies directly below the diaphragm in the upper right hand part

Tip

The appendix can be surgically removed in the event of it becoming infected itself.

Tip

The bend of the colon is known as a flexure. The **hepatic flexure** is the bend between the ascending and transverse colon and the **splenic flexure** is the bend between the transverse and descending colon.

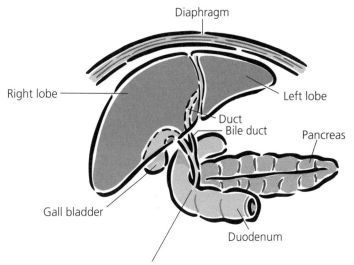

Entry of combined duct into duodenum from pancreas and gall bladder

The liver, gall bladder and pancreas

of the abdominal cavity. It has a large right section and a smaller left section. Each section is known as a *lobe* and the right lobe is connected to the gall bladder by a duct. The liver is one of the most important links between body systems and receives a double supply of blood. It receives oxygenated blood (blood rich in oxygen) via the hepatic artery, which is a branch of the descending aorta, (Chapter 6) and deoxygenated blood together with newly absorbed nutrients via the hepatic portal vein as part of the portal circulation (Chapter 6). As a result, the liver performs many functions not all of which are associated with the digestive system.

- Filtering – the blood from the hepatic portal vein is filtered as it passes through the liver which removes damaged and dead red blood cells and other unwanted particles e.g. excess proteins.

- Detoxification – the liver removes toxins such as drugs and alcohol from the blood.

- Deamination – the liver breaks down the damaged, dead blood cells forming *bilirubin* that helps to form *bile*. It also breaks down other unwanted particles (excess proteins and toxins) forming the waste products **urea** and **uric acid** which are then excreted out of the body in **urine**.

- Storage – the liver stores some vitamins, glycogen and iron from the foods we eat for use by the body at a later date e.g. glycogen for the muscles.

- Production – the liver produces bile which it sends to the gall bladder for storage. It helps to maintain body temperature by producing heat and the break

down of damaged and dead red blood cells in the liver produces waste products.

Gall bladder

The gall bladder is a pear-shaped sac located just above the duodenum and under the liver. It is connected to the liver and the duodenum by ducts. It receives the fluid bile from the liver which it stores until it is needed by the duodenum to aid in the digestive process. Bile consists of water, bile salts to aid digestion and bile pigments e.g. bilirubin which give the faeces its characteristic colour. Gallstones develop as solid particles of bile which can block the passage of bile into the duodenum and cause considerable pain.

The pancreas

The pancreas is a long thin organ lying across the abdominal cavity on the left side of the body. It is also a dual functioning gland comprising of:

- **endocrine** gland meaning that it produces hormones that are secreted directly into the blood stream as part of the endocrine system.
- **exocrine** gland meaning that it produces a fluid substance called pancreatic juice that is passed via ducts to the duodenum and has an effect on the digestive system. Pancreatic juice consists of water, minerals and enzymes.

The digestive system relies on the integration of all its parts and accessory parts to perform its many functions.

Functions of the digestive system

There are four main functions: ingestion, digestion, absorption and elimination.

Ingestion

This involves the *taking in*, *chewing* and *breaking down* of food in the mouth. The food is formed into a soft ball known as a *bolus* by the following structures:

- *Lips* – the nerve supply to the lips judges the temperature of the food and fluid entering the mouth, and the muscular action of the upper and lower lips ensures that it is safely enclosed within the mouth.

Fascinating Fact

Our senses stimulate the salivary glands into producing more saliva. The mere thought, sight, sound and smell of food makes our mouths water and we then need to swallow a lot due to the increased production of saliva in readiness of eating! If we think the food smells and looks bad then the salivary glands produce less saliva and swallowing becomes difficult!

Fascinating Fact

Taste buds are also scattered in the roof of the mouth and the walls of the pharynx enabling us to respond to any unpleasant taste by alerting the body to spit out the offending food or fluid!

- *Teeth* – chisel-shaped incisor teeth bite off large pieces of food; sharp canine teeth tear the food, and molar teeth grind it down.

- *Muscles* – the buccinator muscles draw the cheeks in; the masseter muscles draw the mandible bone up to meet the maxilla bone, exerting pressure on the food in between the teeth; and the temporalis muscles close the mouth.

- *Saliva* – binds and lubricates the food making it easier to swallow. Saliva helps to dissolve the food so that it can be tasted and it also helps to cleanse the mouth and teeth.

- Tongue – tastes the food and fluid as it rolls it around the mouth during the chewing process, before taking it to the back of the mouth once it has formed into a bolus to be swallowed. The taste buds present on the surface of the tongue contain tiny nerves which detect whether or not we wish to continue with the process of eating and drinking by sending a message to the brain which in turn interprets the taste.

- *Pharynx* – the muscles of the pharynx contract and propel the bolus down into the oesophagus. All other areas are closed off as swallowing takes place. The soft palate rises to close off the nasopharynx. The epiglottis closes off entry into the trachea as this clever muscular co-ordination ensures that the bolus goes to the right place.

Digestion

Digestion is the breaking down of food from the pieces we put into our mouths to the minute particles that can be absorbed by cells. Digestion can be divided into two processes:

- Mechanical digestion which involves mastication (chewing) to break down food into a bolus and which takes place in the mouth.

- Chemical digestion which involves the breaking down of food by digestive juices containing enzymes and which takes place in the mouth, the stomach and the duodenum. During this process the bolus becomes liquefied and is known as **chyme.**

- In the mouth saliva is produced by the salivary glands containing the enzyme **amylase**. This starts to digest **carbohydrates**.

Fascinating Fact

The laryngopharynx is primarily a passageway for food and fluid. However, air can also enter the alimentary canal through this part of the pharynx and this can add to the bloated feeling we can experience if we breathe in and swallow at the same time! The air may be propelled back out of the body through the mouth or travel all through the system to be released by the anus. This is commonly referred to as passing wind!

Fascinating Fact

The body produces approximately 5 litres of digestive juices a day. As well as enzymes, these juices contain large amounts of water which continues through the alimentary canal with the chyme.

Fascinating Fact

Some people do not produce enough of certain enzymes for the efficient digestion of certain foods and this can lead to food intolerances which in turn leads to symptoms of bloating, stomach cramps and possibly diarrhoea. Such foods should be avoided.

- In the stomach gastric glands produce gastric juices which contain the enzyme **pepsin**. This is responsible for digesting **proteins**.

- The gastric glands also produce hydrochloric acid which stops the action of the salivary amylase as well as killing off harmful particles that have entered the stomach. When the chyme reaches a certain acidic level within the stomach, the pyloric sphincter allows small amounts to enter into the first part of the small intestine – the duodenum.

- In the duodenum, pancreatic juices from the pancreas enter via the pancreatic duct. The pancreatic juices contain enzymes. These include **lipase** to digest **fat**, **amylase** to carry on the digestion of **carbohydrates** and **trypsin** to carry on the digestion of **proteins**.

- Also in the duodenum, intestinal juices are produced by the villi in the mucosa layer; these contain the enzymes **maltose**, **sucrose** and **lactase** to digest sugar, together with **erepsin** which completes the digestion of **proteins**.

- At the same time, bile produced in the liver and stored in the gall bladder also enters the duodenum. Bile breaks down fats into smaller particles in a process called **emulsification**.

As a result of the digestive process, the food we eat goes through a series of changes from the large particles we put into our mouths into the easy-to-swallow bolus and the liquefied chyme. Carbohydrates, proteins and fats need to be broken down further by enzymes so that the next process can take place. Vitamins, minerals and water do not need to be broken down further as they are made up of small enough particles.

Absorption

Absorption is the process of passing nutrients from the alimentary canal into the circulatory system for use around the body. Absorption takes place in the stomach, small intestine and large intestine.

- Stomach – limited amounts of water, alcohol and some drugs are absorbed directly into the bloodstream and circulated around the body.

- Small intestine – peristalsis muscular action pushes the chyme through the duodenum, jejunum and ileum and the villi in the mucosa layer allow the absorption of digested nutrients to take place.

Present in the villi are blood capillaries which absorb into the blood stream the digested carbohydrates, proteins, vitamins, minerals and water. Also present in the villi are lymphatic capillaries called lacteals which absorb the digested fats before passing them into the blood stream. The blood distributes the nutrients to the cells as required on its way around the body and stops off at the liver to be filtered and detoxified and to drop off excess nutrients for storage. As the chyme reaches the end of the small intestine, most of the nutrients have been absorbed by the blood and lymph leaving indigestible food, water and a few nutrients.

- Large intestine – as the chyme reaches the ileum at the end of the small intestine, the ileocaecal sphincter opens to allow entry into the large intestine and closes again to prevent back flow. Any remaining nutrients are absorbed and the 'leftovers' form into **faeces**. Peristaltic action forces the faeces along the rectangular colon to the rectum. Any remaining water is absorbed en route.

Elimination

This involves the release of indigestible food from the body as waste.

- When the faeces reach the rectum, it triggers a reflex action producing a sensation which we recognise as the need to defecate. Peristaltic waves force the faeces along the anal canal and the internal sphincter relaxes. The external sphincter is under voluntary control and at this stage we can either 'go with the flow' or alternatively squeeze the muscle tightly to prevent defecation and override the sensation until a more convenient time or place!

This whole process takes place over a period of hours and sometimes days depending on how difficult it is for the processes to take place. Rich, tough foods take longer to digest and will stay in the stomach longer than easily digested, softer food. Absorption of nutrients takes place over the following few hours with elimination usually taking place several hours after that. All these processes take place more efficiently if the body is not involved in excessive activity. The digestive system likes relaxation time when the blood can be directed to its various parts and away from the muscles, and it is for this reason that we feel lethargic after eating and suffer with the symptoms of indigestion if we remain excessively active.

Common conditions

An A–Z of common conditions affecting the digestive system

- ACID REGURGITATION – a condition in which the contents of the stomach along with the hydrochloric acid from the gastric juices re-enter the oesophagus causing it to become inflamed.

- ANOREXIA – avoidance of eating resulting in extreme deficiency and in severe cases death by starvation. It is a psychological disorder which involves the fear of eating for any number of reasons.

- APPENDICITIS – inflammation of the appendix. Acute appendicitis strikes suddenly and results in an operation to remove the appendix. Chronic appendicitis can continue for several months without the need for surgery.

- BOWEL CANCER – cancer affecting the large intestine. Can affect any part of the large intestine with growths that constrict the passageways.

- BULIMIA – a disorder associated with over-eating followed by self-inflicted vomiting and/or the use of laxatives. As with anorexia, bulimia is a psychological disorder and normal eating habits will only be resumed after the problem has been resolved.

- CIRRHOSIS OF THE LIVER – hardening of the liver generally caused by the consumption of excessive amounts of alcohol.

- COELIAC DISEASE – an intolerance of gluten (a protein found in wheat).

- COLITIS – an inflammation of the large intestine resulting in diarrhoea which is then combined with blood and mucus as a result of damage to the intestine lining.

- CONSTIPATION – infrequent passing of waste from the anus resulting in dry, hard faeces as much of the water is absorbed making it increasingly difficult to pass.

- CROHN'S DISEASE – see ILEITIS.

- DIARRHOEA – abnormally frequent elimination of faeces during a peristaltic 'rush' resulting in dehydration and weakness, as too much fluid and too many nutrients are lost from the body too quickly.

- DYSENTRY – an infection of the large intestine causing severe diarrhoea.
- FLATULENCE – air in the stomach or intestines that has been swallowed whilst eating or drinking. It may also be associated with certain foods, which produce a gas as digestion takes place.
- GALLSTONES – solid mass formation of particles of bile found within the gall bladder causing a possible obstruction and preventing bile from passing into the duodenum. Extreme cases may lead to the complete removal of the gall bladder.
- GASTRITIS – irritation or inflammation of the stomach. Can be associated with something that has been eaten or drunk.
- GASTROENTERITIS – inflammation of the stomach and intestines resulting in vomiting and diarrhoea. Dehydration and weakness can occur very quickly so care must be taken to replace lost fluid and nutrients.
- HERNIA – a rupture (tear) whereby an organ penetrates its protective covering. A common site of a hernia is the large intestine in men.
- HICCUPS – repeated and involuntary spasms of the diaphragm.
- INDIGESTION – pain associated with eating certain foods which are more difficult to digest. It is also associated with over-eating, hunger and other often stress-related disorders.
- ILEITIS – an inflammation of the latter section of the small intestine (ileum). Also known as Crohn's disease.
- IRRITABLE BOWEL SYDROME (IBS) – a condition commonly associated with high stress factors. Symptoms include alternate bouts of diarrhoea and constipation.
- JAUNDICE – a yellow discolouration of the skin which signifies serious disease in adults. The yellow colour comes from bilirubin which is formed as a result of the breaking down of old blood cells in the spleen.
- OBESITY – excessive weight gain through over-eating.
- OESOPHAGITIS – inflammation of the oesophagus often associated with heartburn (a burning sensation in the chest).
- OESOPHAGUS CANCER – malignant growths along the oesophagus. Most commonly found at the

lower section of the oesophagus and occurs mainly in middle-aged men.

- PILES – swollen veins in the anus causing pain and discomfort. Bleeding from these veins can result in anaemia due to the excessive loss of iron in the blood.

- PROCTITIS – inflammation of the lining of the rectum resulting in pain while passing faeces together with the urge to pass more.

- PROLAPSE – the displacement of a part of the body e.g. rectum.

- STOMACH CANCER – a malignant disease that is more common in men over the age of forty.

- ULCERS – a break in the surface of any part of the body. Commonly associated with various parts of the digestive system where there is a break in the lining of the alimentary canal caused by the over-production of acid in the gastric and intestinal juices.

System sorter

DIGESTIVE SYSTEM

Skeletal

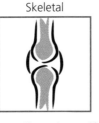

The maxilla and mandible bones hold the teeth which assist in the ingestion of food in the mouth.

Muscular

The peristaltic action of the smooth layer of muscle in the alimentary canal propels the food we eat through the digestive system. These muscles are involuntary.

Integumentary

The skin produces vitamin D which helps in the absorption of calcium in the small intestine. Calcuim is found in many foods consumed as part of our diet.

Respiratory

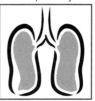

Oxygen from the lungs activates glycogen from the digestive system to produce energy in the cells.

The blood transports nutrients and water from the digestive system to the cells of the body. Lymphatic capillaries and lacteals absorb fats from the small intestine before passing them on to the blood for distribution

Circulatory

Excess water and minerals from digested food are formed into urine in the kidneys and stored in the bladder until released from the body.

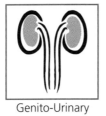

Genito-Urinary

During a stressful situation the hormone adrenalin will cause the digestive system to shut down temporarily by directing blood to the muscles. Therefore long term stress can have a detrimental effect on the digestive system.

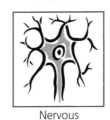

The nervous system is responsible for initiating the feelings of hunger alerting a person to eat. It also alerts a person to the urge to release the waste from the anus.

Nervous

Endocrine

The digestive system is made up of the alimentary canal with associated organs which are involved in:

- ingestion of food in the mouth
- digestion of food in the stomach
- absorption of nutrients in the small intestine
- elimination of waste in the large intestine

providing the whole body with the vital 'fuel' to perform their various functions. In turn the digestive system relies on its links with the other systems.

Holistic harmony

The efficient functioning of the digestive system ensures that the cells, tissues, organs and systems of the body are provided with the nutrients and water necessary for their survival. The digestive system also relies on itself and its links with the other systems for its own care.

Fluid

The body loses about one and a half litres of water a day through the kidneys as urine, through the lungs as we breathe out, as sweat through the skin and in faeces. The body makes about a third of a litre of water a day through the processes of energy production in the cells. Therefore the body needs a minimum intake of just over a litre of water a day in order to maintain its fluid balance and to avoid the problems associated with dehydration. Drinking water aids the digestive system by preventing constipation from occurring i.e. if the urge to defecate is resisted over a period of time, the faeces begin to dry out as more of the water is re-absorbed into the blood. This makes the passing of faeces increasingly difficult and painful and may involve the straining of the lower digestive system. Constipation has a 'knock on' effect on the other body systems contributing to the sluggish appearance of skin when the toxins held within the faeces are kept in the body longer than they are naturally intended.

Nutrition

The digestive system is responsible for breaking down the elements of the food we eat so that absorption of the nutrients can take place which is part of the natural process of keeping the body alive. The food we eat can be classified as:

 Fascinating Fact

Sugar diabetes is caused by high levels of glucose in the blood caused by a lack of effective hormones (Chapter 10).

1. Carbohydrates – which are broken down into glucose and taken by the blood to the liver. The liver sends some glucose to the muscles to be stored as muscle glycogen, which is then used with oxygen for the production of energy. The liver also stores some of the glucose itself in the form of glycogen to send to the muscles at a later date, and the rest of the glucose is circulated by the blood to be distributed to the cells with any excess being converted into fat for storage. There are 'fast releasing' carbohydrates such as sugar, sweets and most fast foods which provide us with a 'quick fix' of energy and slow

releasing carbohydrates such as whole grains, vegetables and fresh fruit which provide us with a more sustained flow of energy.

2. Proteins – which are broken down into smaller particles called amino acids and which provide the body with the means for growth and repair. The proteins that we eat in the form of eggs, cheese, meat, fish, soya, lentils, peas and beans etc. are broken down into different types of amino acids through the digestive process. These amino acids are then absorbed by the blood and taken to the liver where they are either sent out to be used by the cells of the body; used by the liver cells to form plasma proteins; transaminated (changed from one type to another); or deaminated (those that are not required are broken down further and formed into the waste product urea which is taken to the kidneys by the blood and eliminated out of the body in the form of urine).

3. Fats – these are absorbed into the lymphatic system via lacteals during the digestive process of emulsification, before entering the blood flow at the lymphatic ducts; they provide the body with another source of fuel. Fats are used to make parts of cells and to provide energy. Any excess fat is removed from the blood and stored until required. There are two main sources of fats: saturated fats or hard fats from dairy products and meat and unsaturated fats or soft fats from vegetables, nuts and fish. Saturated fats are found in many processed foods and are not as useful to the body as unsaturated fats.

4. Vitamins – A, B, C, D, E and K are absorbed by the blood from the digested foods from which they originate and they provide help with every process within the body. Excess vitamins can be stored in the body to be called upon in the event of deficiency which may occur for example during dieting. Vitamins A and B12 are stored in the liver and vitamins A, D, E and K, which are fat soluble, are stored in the fat cells.

5. Minerals – such as iron, calcium, sodium, chloride, potassium, phosphorus, magnesium, fluoride, zinc, selenium etc. are absorbed in much the same way as vitamins and are also necessary to assist the various processes that take place within the body. Surplus minerals are either not absorbed so lost from the body in faeces, or taken to the kidneys and lost through urine excretion.

6. Fibre – is tough fibrous carbohydrate that cannot be digested. Insoluble fibre, such as cellulose found in wheat bran, fruit and vegetables, make it easier for

Remember

Remember that any excess fat is stored in the body as fatty tissue forming the subcutaneous layer which lies between the dermis and the muscles. This fatty tissue may be used for energy in the event of lack of food e.g. dieting.

Fascinating Fact

Cholesterol forms the waxy substance responsible for blocking arteries. It is essential in the body for the formation of cells and hormones and is formed in the liver as well as being present in our diet. Saturated fats contain cholesterol and eating an excess of this type of fat is one of the causes of heart attacks due to the clogging of vital arteries. It also contributes to the formation of gallstones in the gallbladder.

faeces to pass along the large intestine by adding bulk. This bulk absorbs water making the faeces soft. The muscular layer of the large intestine is stimulated by the bulk and as a result the waste leaves the body more quickly, reducing the risk of constipation and infection.

It is clear that the body needs a balance of nutrients to aid in the functions of the digestive system and in turn the functions of the body as whole. Ignoring the need to eat immediately affects the short-term running of the body and results in dehydration and weakness; in time it will affect the long-term efficiency of the body resulting in illness and eventual death.

Rest

The body needs to rest to allow the digestive system to process the food we eat. During and directly after eating the body needs a short resting period to allow the alimentary canal to perform its various functions. Large amounts of blood are required by the digestive system to allow this to happen naturally and efficiently. Blood is diverted from other systems of the body to the digestive system during periods of inactivity. However if the body remains active during or directly after eating, the blood is needed to sustain that activity taking it away from the process of digestion. When this occurs, the body experiences problems associated with poor digestion resulting in feelings of nausea, bloating, flatulence and indigestion. Rest also provides the body with the time it needs to absorb the food we have eaten so that the nutrients are available for use when they are needed. Subsequently, the elimination of waste occurs much more efficiently after we have rested.

Activity

Activity is possible when food and fluid have been ingested, digested and absorbed. Carbohydrates, proteins and fats that have been eaten are broken down during digestion to be used as energy when they are absorbed by the blood and taken to the cell (cellular metabolism). When the body experiences a lack of food it will draw on its reserves from the muscles, liver and fat cells. Eating more than we need for energy will result in weight gain in the same way as eating less than we need for energy will result in eventual weight loss. Energy value in food is measured in kilocalories (kcal) or kilojoules (kJ). 1 kcal = 4.2 kJ and the average requirement for adults is approximately 1940 kcal/8100kJ for a woman and 2550 kcal/10,600kJ for a man. In order to maintain body weight, we should aim to consume the amount of food needed to supply the body with just the right amount of energy. The amount of energy needed per person per day fluctuates depending on age, sex, size, and levels of activity; it will also change during pregnancy, breast-feeding and illness. The body is alerted to the need to increase energy levels with feelings of hunger. However we are often confused by these feelings, as we are tempted to eat for other reasons such as boredom, habit, sociability or just because the food is there! We are also good at ignoring the signs that we have eaten enough and often find ourselves over indulging!

Air

Air inhaled from the atmosphere contains the life force (oxygen) needed to activate the energy taken into the body by eating. The way in which we breathe has an impact on the amount of energy activated at any one time and should be adapted to meet the needs of the body. Our breathing quickens when more energy is needed and slows down considerably when less energy is required. It is important to slow our breathing down when we are eating to avoid taking excessive air into the digestive system, and to increase our breathing when we want to use the energy gained from food by activating it with an increased supply of oxygen. Although breathing is an involuntary action controlled by the respiratory (Chapter 5) and nervous systems (Chapter 9), we can exert an element of control over how we actually breathe. If more attention was paid to the art of breathing, the body would experience far less general stress and trauma which in turn would help to prevent the onset of certain illnesses as well as help to alleviate their symptoms e.g. irritable bowel syndrome can be helped by correct breathing.

Age

Age affects the energy requirements of the body, with children requiring greater levels than the elderly. As we age the body slows down and this is reflected in the amount of food needed to sustain the decreasing levels of activity. Middle age is often accompanied by weight gain because we ignore the need to lessen our intake of food and eating habits can be hard to change, especially if we associate eating with pleasure. In addition, ageing affects the digestion and absorption functions of the system. The production of enzymes declines with age, therefore the digestion of food is impeded which leads to poor absorption of nutrients.

Colour

The alimentary canal covers a large part of the body starting at the mouth and descending down to the anus. It passes through five of the chakras starting at the fifth and ending at the first. As a result, the digestive system is affected by the array of colours associated with each of the chakras it passes through.

- Blue of the fifth chakra assists the throat generally.
- Green of the fourth chakra helps to balance the system.
- Yellow of the third chakra is thought to purify as it has an effect on the stomach, liver, pancreas and small intestines helping the digestion and absorption of nutrients.
- Orange of the second chakra continues the cleansing process and assists the passage of waste through the small and large intestines.
- Red of the first chakra helps the elimination process and prevents the lower digestive system from becoming sluggish.

The use of colour is often an intuitive act and you may find yourself choosing an outfit in a colour that will have a positive effect on a stressed part of your body!

 Angel advice

We should learn to trust our instinct and encourage our children not to lose sight of it in the first place!

Awareness

Awareness of the way in which the digestive system contributes to the well-being of the whole body provides the key to healthy eating. This is aided further by the creation of balance between the physical and psychological need for food when we take note of the signals put out by the body. Children have an inbuilt awareness of exactly what is needed and when, and if left to their own devices with adequate access to food and fluid, a child would neither starve nor over eat. We quickly lose this intuitive instinct to eat and drink as we become bound by the rules of society – rules which were made with no real thought to the needs of the digestive system! What sense is there in missing breakfast in the morning when we need maximum nutrients to see us through the day, and indulging in a three-course meal at the end of the day when we are not going to need any energy for at least twelve hours?

Special Care

The care given to the digestive system is reflected in the general health of a person. If the digestive system has been cared for it will in turn care for the whole body. The digestive system processes the 'fuel' needed to sustain activity and so the quality and quantity of 'fuel' should be considered together with the time it takes the system to ingest, digest and absorb. Stress plays a part in upsetting the equilibrium needed for the efficient processing of this 'fuel' and is one of the main causes of digestive problems. Stress affects the digestive system by effectively 'switching it off' until the stressful situation has passed. In addition to this, stress plays a part in how we feel about food. Some people will turn to food for comfort when they are stressed and experience a rapid gain in weight, whilst other people turn away from food when faced with a stressful situation and have the opposite problem of extreme weight loss. Care of the digestive system should involve:

- Regular meals to provide enough energy to sustain activity
- A balanced diet to maintain the healthy functioning of the whole body
- At least one litre of water per day to avoid dehydration
- Fresh, unprocessed food for maximum nutritional value
- Taking time out to eat to avoid indigestion
- Taking time out to visit the toilet for efficient, regular elimination
- Avoiding intense activity during and directly after eating
- Eating when hungry not out of habit or boredom
- Chewing food well to assist in mechanical digestion
- Avoiding excessive stressful situations which can have a detrimental effect on digestion, absorption and elimination
- Avoiding sources of free radicals such as fried or browned food which contribute to premature ageing.

Think about the times you rush your food, eat on the run or even skip meals and then over-indulge in highly processed fast foods because you are hungry but too stressed, lazy or busy to prepare a proper meal. It's no wonder that digestive problems are so common!

Angel advice

Learning about the digestive system should help us to take that extra little bit of care needed to allow the system to do what it should, when it should for the good of the whole body. Try it and see – it's easy when you know how, and more importantly, why!

Treatment tracker

DIGESTIVE SYSTEM

Make up

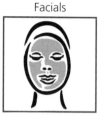

Shading techniques may be used to hide fatty tissue associated with overweight and define areas associated with underweight.

Facials

Facial massage encourages deep relaxation which inhibits the flow of adrenalin. This relaxation assists the efficient working of the digestive system.

Nail care

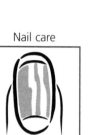

Hand and foot massage speeds up the use of nutrients in the cells of the skin and nails (cell metabolism). This improves the overall condition, growth and repair.

Hair removal

The removal of hair stimulates blood flow to the surrounding area of skin speeding up the use of nutrients (cellular metabolism) from the digestive system aiding growth and repair.

Massage along the large intestine can successfully ease the symptoms of constipation by aiding peristaltic action.

Massage

Reflexology may be used to ease the symptoms associated with IBS by creating balance between the digestive system and the other systems of the body

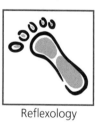

Reflexology

Essential oils suitable for the digestive system include peppermint for heartburn, bloating, diarrhoea and nausea. Black pepper for constipation and loss of appetite.

Aromatherapy

Excessive nutrients eg fats and carbohydrates are stored as fatty tissue which may be worked on with the use of electrical treatments in order to break them down as part of a calorie controlled diet.

Electrical

The digestive system responds well to the relaxation effects of beauty and holistic therapies, helping to balance the activity within the body between all of the systems.

Knowledge review – Digestive system

1 Name the main parts of the alimentary canal and the three accessory organs.

2 What is the term given to the food that has been broken down in the mouth and is ready to be swallowed?

3 What mixes with the food in the mouth?

4 Where does the muscular co-ordination take place that ensures that the food enters the digestive system and not the respiratory system?

5 What is the term given to the muscular action which moves the food along the alimentary canal?

6 What is the tube that links the pharynx with the stomach?

7 What are the protein catalysts called that assist in the digestive process?

8 What is chyme and where is it formed?

9 Name the parts of the small intestine.

10 Where is bile formed, stored and secreted?

11 How does the pancreas assist with digestion?

12 What does the liver store?

13 What does the liver do to blood?

14 What nutrients are needed to make up a well-balanced diet?

15 What structures are responsible for the absorption of nutrients in the small intestine?

16 What are faeces made up of?

17 Name the main parts of the large intestine.

18 Where does elimination of faeces take place?

19 What are the appendix, tonsils and adenoids made up of and what is their function?

20 How can massage benefit a person suffering from constipation?

The genito-urinary system

Learning objectives

After reading this chapter you should be able to:

- **Recognise the structures that make up the genito-urinary system**

- **Identify the structure of the kidneys and their effect on the body's fluid balance**

- **Understand the functions of the genito-urinary system**

- **Be aware of the factors that affect the well-being of the genito-urinary system**

- **Appreciate the ways in which the genito-urinary system works with the other systems of the body to maintain homeostasis.**

Our journey through the human body continues with input and output, focusing on two systems that are closely linked together in terms of organs – the reproductive system which contains the organs of reproduction – and the urinary system which contains the organs responsible for the elimination of excess water and waste from the body. Some of the organs share a dual function and so are often classified as a combined system – **THE GENITO-URINARY SYSTEM**. The reproductive organs are different in a male and female although their prime functions are the same, that of the reproduction of a new human being. The organs of the urinary system (also known as the renal system) are similar in a male and female and their functions contribute to the balance of water levels within the body.

Science scene

Structure of the male and female reproductive organs

The genito-urinary system consists of the male and female **genitalia**, the **kidneys**, **bladder**, **ureter** tubes leading from the kidneys to the bladder and a **urethra** tube leading from the bladder to the outside of the body.

Genitalia

The main organs of reproduction are located within the pelvic girdle which forms a protective casing of bone (Chapter 3) in the lower trunk.

The male genitalia include the **testes**, the **vas deferens**, the **prostate gland** and the **penis**. The female genitalia include the **ovaries**, the **fallopian tubes**, the **uterus** and the **vagina**.

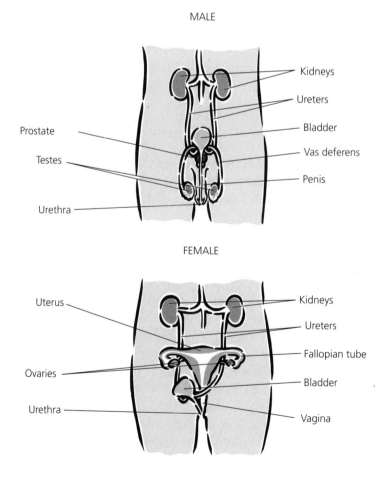

The male and female genito-urinary systems

Male genitalia

There are two testes which develop in the abdomen of the male foetus and drop into a sac of skin and muscle located behind the penis known as the **scrotum** just before birth. After puberty, the testes start to produce **spermatozoa** (sperm) which is passed from the testes through a tightly coiled tube called the **epididymis**. Each epididymis is about six metres long and connected to the testes via ducts. The sperm produced in the testes are immature and develop as they pass through the epididymis. The epididymis develops into a larger tube approximately 45 cm in length, known as the vas deferens, which creates a passageway for the sperm from the testes to the penis. Peristaltic movements move the sperm along. Each vas deferens passes by sac-like structures called the **seminal vesicles** which secrete part of the fluid that mixes with the sperm to form the liquid **semen**. This fluid is passed into the vas deferens through ducts. Each vas deferens then passes through the prostate gland, a chestnut-shaped structure, where it picks up a milky fluid which also contributes to the formation of semen, before joining the urethra tube to create an exit out of the body through the penis.

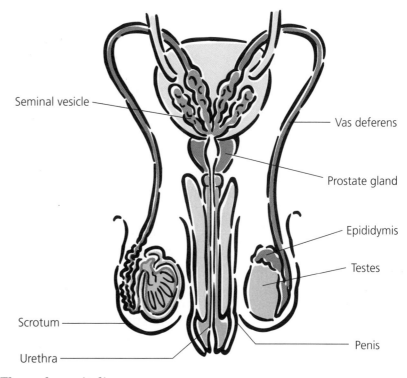

The male genitalia

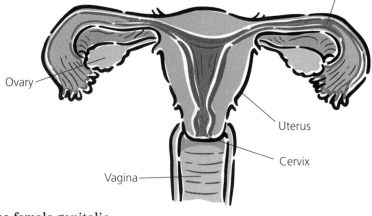

The female genitalia

Female genitalia

There are two ovaries located on either side of the pelvic girdle, about the size and shape of almonds, which store **ova** (eggs). The fallopian tubes, each approximately 10 cm in length, open just above the ovaries creating a passageway from the ovaries to the uterus which is located in the centre of the pelvic girdle.

Each month after puberty an ovum (single egg) is released from one or other of the ovaries and passes along the corresponding fallopian tube to the uterus. The uterus is often referred to as the **womb** and is the site for the development of an ovum which is fertilised by a sperm or the release of the unfertilised ovum. The uterus opens out into the vagina at the **cervix** or neck of the womb which together form the route for the birth of a baby or the blood associated with menstruation.

Accessory organs

The breasts are glands which develop in females in response to activity within the genitalia. They are present in males but are not activated into development in the same way.

The female breasts are known as **mammary glands**. They develop during puberty and contain connective tissue in the form of adipose and areolar tissue. The pectoralis major muscles (Chapter 4) lie directly below the breasts and strands of connective tissue called **Cooper's ligaments** help to support them. Each breast consists of about twenty lobes. Each lobe contains glands called **alveolar glands** and an *alveolar duct*. Milk is produced in the glands in response to pregnancy. It is passed through the duct to the **lactiferous sinuses** where it is stored. During breast-feeding, the milk passes into further ducts and out of the

Fascinating Fact

A female holds in the region of 30,000 undeveloped ova from birth!

Fascinating Fact

The cervix is very narrow in diameter but is able to dilate (open) to approximately 10 cm to allow for the birth of a baby.

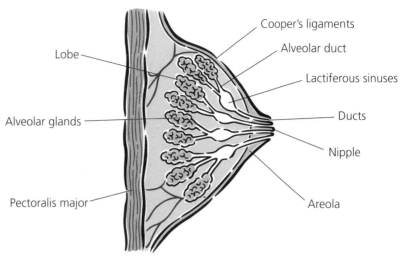

The female breast

nipple, which forms the external portion of the breast. Around the nipple, the pigmented skin is called the **areola** and contains sebaceous glands that secrete sebum to lubricate the area.

The following structures are common in both sexes and form the urinary or renal system.

Kidneys

There are two kidneys, each about the size of a tightly clenched fist. They are bean shaped and situated either side of the spine at waistline. The right kidney lies slightly

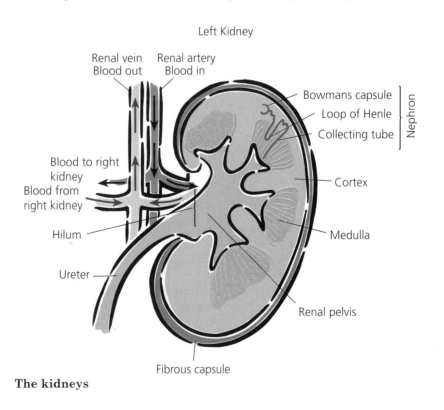

The kidneys

lower than the left because of the restricted space caused by the position of the liver above. Each kidney is protected by the lower ribs, abdominal wall, fatty tissue and muscles of the back and consists of an outer fibrous capsule providing further protection.

The concave inner section of each kidney is known as the **hilum** which opens into the **renal pelvis,** which in turn leads to the inner **medulla** and finally, the outer **cortex**. Situated within the medulla and cortex are small tubes called **nephrons**. There are approximately one million nephrons in each kidney which loop in and out of the cortex and medulla before entering the renal pelvis. They begin as blind-ended tubes in the cortex where they form into cup-shapes called **Bowman's capsules**. They then extend down into the medulla forming a **loop of Henle** before returning into the cortex again. The nephron re-enters the medulla, this time forming into a collecting tube which passes into the renal pelvis and hilum.

The hilum provides entry *into* the kidneys for the arterial blood via the renal artery, which sub divides into afferent arterioles and finally tiny capillaries forming a tight knot called the **glomerulus** situated within the Bowman's capsule. Efferent arterioles take the blood from the glomerulus uniting with the renal vein at the hilum via capillaries to *exit* the kidneys. The hilum also provides entry *out* of the kidneys for the ureters, which lead from the renal pelvis to the bladder.

Ureters

The ureters are two long, narrow muscular tubes leading from the kidneys to the bladder. They are approximately 25 cm in length and only 3 mm wide. They are made up of three layers:

1. Outer fibrous layer which is continuous with the fibrous capsule of the kidneys

2. Muscular middle layer consisting of interlacing smooth, involuntary muscle

3. Inner layer of epithelial cells.

Bladder

When empty the bladder is pear-shaped, becoming oval in shape as it fills with urine. It is made up of a modified version of the ureters. The inner layer consists of folds called **rugae** which extend as the bladder fills.

Nerve endings located in the wall of the bladder detect any increase in bladder size and activate an internal sphincter

Remember

Remember the rugae of the stomach? The bladder and stomach are similar in that they both need to extend when full.

muscle which relaxes as the bladder contracts, allowing a passageway through into the urethra – the final section of the renal system.

Urethra

The urethra is a single narrow muscular tube which leads from the bladder to the outside of the body. There is an external sphincter muscle situated at the very end of the urethra. When relaxed, this muscle allows entry out of the body, and entry is closed off when the muscle is contracted. This sphincter muscle is made up of skeletal muscular tissue and is therefore under voluntary control.

The urethra is shorter in women than in men – approximately 4 cm in a woman and 20 cm in a man. In a male, the vas deferens joins the urethra, which then provides a dual passageway for two types of fluid, either semen or urine, out from the penis.

In a female, the urethra is responsible for carrying just urine.

Functions of the genito-urinary system

The functions of the genitalia are the production of hormones and reproduction.

There are five functions of the urinary system: filtration, re-absorption, production, excretion and regulation.

Production of hormones

The testes in males and the ovaries in females are also known as the **gonads** or sex glands and are responsible for producing the male and female sex hormones. The testes produce male hormones known as **androgens** which include **testosterone,** and the ovaries are responsible for producing the female hormones **oestrogen** and **progesterone** (Chapter 10).

In males these hormones are responsible for producing the sexual characteristics associated with the onset of **puberty** i.e. the development of sperm, the change of voice and the growth of facial hair etc. In females, they are responsible for the onset of **puberty** and the start of the **menstrual cycle**, the temporary and final cessation of the menstrual cycle during pregnancy and eventually the **menopause**.

Remember

The urethra can only allow one type of fluid out at any one time.

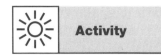

Activity

If an ovum is not fertilised in this way, it is released from the body and a new ovum will be formed in the ovaries as part of the menstrual cycle.

Remember

Remember that afferent means to arrive and efferent means to exit.

Reproduction

The onset of puberty equips a male and a female to reproduce a new human being and is a process which ends in a female during the menopause and goes on almost indefinitely in a male. Reproduction is reliant on healthy sperm and ova, and during menopause the ovaries stop developing a new ovum each month although a male of the same age will still be capable of producing healthy sperm.

Reproduction begins with *fertilisation*. Semen from the male is introduced into the female via the penis and the vagina during sexual intercourse. The sperm present in the semen travel up into the uterus and the fallopian tubes. If an ovum is present, the sperm will try to penetrate it so that fertilisation can take place. If it is successful, the nucleus of the sperm fuses with the nucleus of the ovum forming a **zygote**. The zygote contains all the elements (chromosomes) needed to form a new human being, 23 from the sperm and 23 from the ovum. This process of reproduction is known as **meiosis** (Chapter 1). The zygote undergoes a series of simple cell divisions known as **mitosis** (Chapter 1) until an **embryo** is formed which develops further in the uterus until it becomes a **foetus** and finally a baby. This whole process takes approximately forty weeks.

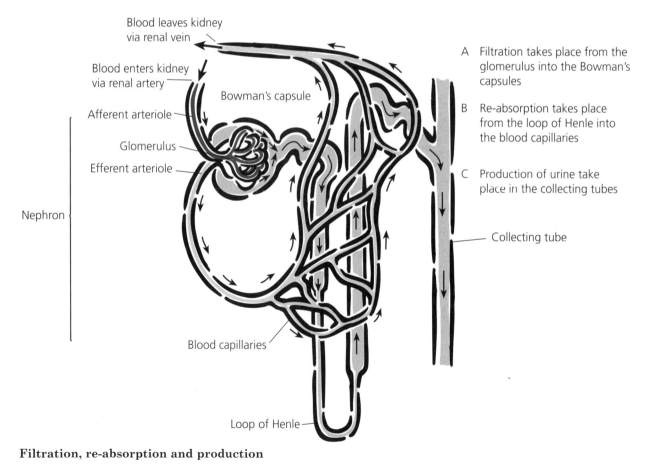

Filtration, re-absorption and production

Filtration

Approximately one pint of blood passes through the kidneys every minute for filtration, as follows:

1. The blood enters the kidneys via the renal artery through the hilum where it subdivides into afferent arterioles.

2. These arterioles subdivide further in the cortex into groups of tiny capillaries forming the glomerulus.

3. The nephrons surround the glomerulus at the Bowman's capsules.

4. The blood passes some substances out from the glomerulus and into the Bowman's capsules. These substances include:
   - Water, minerals, glucose, amino acids and vitamins which have been absorbed into the blood from the digestive system.
   - Urea and uric acid which are formed in the liver from the deamination of excess amino acids and worn out cells.
   - Hormones, drugs and toxins.

5. The blood then leaves each glomerulus via efferent arterioles which form into tiny capillaries surrounding the rest of the nephron for the next stage of the process.

Re-absorption

As the substances pass through the nephron into the loop of Henle, all of the glucose, amino acids and vitamins are re-absorbed into the blood capillaries together with some of the minerals and most of the water. The unwanted water, minerals and waste are left behind in the nephron. The blood capillaries all unite with the renal vein at the hilum as the blood leaves the kidneys to carry on circulating around the body.

Production

Production of urine takes place as the unwanted water, minerals, and waste including urea, uric acid, hormones, drugs and toxins pass along the nephron into the collecting tube.

The urine is then taken to the renal pelvis. From here it passes out of the kidneys through the hilum and into the ureters, where peristaltic contractions move it along to the bladder for storage.

Remember

Recall the action of the anus in the digestive system – it also has internal and external sphincter muscles which control the elimination of faeces in the same way.

Fascinating Fact

Sometimes harmless substances that have no value to the body are injected into the blood to see how efficient the renal system is. If the kidneys are functioning well, the unwanted substances will be found in the urine!

Fascinating Fact

At high altitudes where the levels of oxygen are low, the kidneys respond to the drop in oxygen levels by alerting the bone marrow to produce more red blood cells. Remember that red blood cells absorb oxygen – if there are more cells then more oxygen can be absorbed. The kidneys will halt production when normal oxygen levels are restored.

The production of urine is increased when we drink more and decreased as we drink less and/or lose water through sweating. The sweat glands are responsible for excreting water and waste from the body in response to changes in temperature (Chapter 2). As the body temperature rises, sweat is produced to cool the body which in turn slows down the production of urine in an attempt to prevent dehydration from occurring.

Excretion

The elimination of urine occurs when the bladder becomes full. The bladder wall contracts and the urine passes through the relaxed internal sphincter and into the urethra. At the same time, the nerve endings in the bladder wall inform the brain which in turn alerts the body of the urge to urinate. The voluntary action of the external sphincter overrides this urge and we are able to 'hold off' the need to urinate until a more convenient time and place. The technical term used for the passing of urine is **micturition**.

Regulation

Regulation of water and minerals, the rate of production of red blood cells and maintenance of blood pressure, take place in the kidneys:

- Hormones are released into the blood from the endocrine system and inform the kidneys to either absorb more water and minerals if the levels in the body are low or alternatively produce more urine if the levels are high.
- The kidneys help to maintain the rate of formation of red blood cells by producing a hormone which travels to the bone marrow via the blood to stimulate cell production when levels are low.
- The kidneys produce the enzyme **rennin** which helps in the regulation of blood pressure.

Common conditions

An A–Z of common conditions affecting the genito-urinary system

- AMENORRHOEA – absence or abnormal stoppage of menstruation caused by an imbalance of hormones.
- CANCER – the development of malignant cells e.g. breast cancer, ovarian cancer, bladder cancer etc.
- CERVICAL EROSION – changes in the lining of the neck of the womb accompanied by a slight discharge.
- CYSTITIS – inflammation of the bladder resulting in a frequent urge to pass urine with subsequently small quantities of urine being passed often, accompanied by stinging and pain in the lower abdomen. A common complaint which usually affects women more than men due to the difference in the size of the urethra. In women, the urethra is shorter and bacteria can enter the bladder more easily.
- DYSMENORRHOEA – painful periods caused by hormone imbalance.
- ECTOPIC PREGNANCY – occurs when the fertilised ovum develops outside of the uterus, usually in the fallopian tubes.
- ENDOMETRIOSIS – cells from the uterus form in the fallopian tubes or ovaries causing pain especially during menstruation.
- FIBROIDS – non-cancerous growths which develop in the walls of the uterus.
- KIDNEY FAILURE – may occur if both kidneys become seriously diseased. Kidney dialysis can be used to perform the functions of the kidneys. A dialysis machine acts as a replacement kidney, filtering the blood of unwanted substances.
- KIDNEY STONES – formed by an excess of salts in the blood. These salts crystallise in the urine and can cause an obstruction to the flow of urine resulting in extreme pain. Large stones may be surgically removed while small stones pass out of the body in urine.
- MASTITIS – tender lumps present in the breasts of females associated with the lead up to menstruation.
- NEPHRITIS – an inflammation of the glomerulus inside the kidneys. Nephritis can be either acute or chronic. In cases of acute nephritis, the symptoms

last for two to three weeks often following an acute throat infection. Chronic nephritis may lead to kidney failure.

- PRE-MENSTRUAL TENSION (PMT) – physical and psychological tension leading up to menstruation caused by an imbalance in hormones.

- PROSTATITIS – inflammation of the prostate gland.

- PYELONEPHRITIS – inflammation of the kidneys due to an infection. There are two types – acute which may last a few days, and chronic which can last for many years and if left untreated, can cause kidney failure and death.

- SEXUALLY TRANSMITTED DISEASES – a variety of diseases acquired through sexual contact.

- THRUSH – an infection of the vagina. Symptoms include itchiness and discharge.

- URETHRITIS – inflammation of the urethra. Acute urethritis may last a few days with chronic urethritis lasting for years.

System sorter

GENITO-URINARY SYSTEM

Skeletal

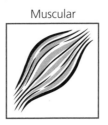

Muscular

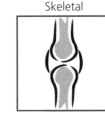

Integumentary

The kidneys and the bones help to control the amount of calcium in the blood by storing some in the bones and excreting some from the body in urine.

Smooth muscular tissue is responsible for the passage of

- urine through the system
- sperm from the testes to the urethra
- ova from the ovaries to the vagina.

Respiratory

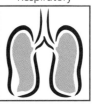

When the skin loses excess water through sweating, the kidneys compensate by releasing less water in urine helping to maintain the body's fluid balance.

Breathing aids the genito-urinary system as the physical action gently massages the organs. The kidneys help to balance any water lost through breathing.

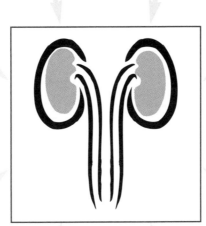

The blood carries urea, excess water and minerals to the kidneys to be made into urine and passed out of the body via the urethra.

Maintaining the fluid balance within the body helps avoid the problems associated with constipation keeping the digestive system clear.

The nervous system is responsible for alerting the body to the need to pass urine and activates the release of urine from the bladder when it is full.

The testes in men and the ovaries in women have an endocrine function. They produce the male and female hormones responsible for secondary sexual characteristics.

Circulatory

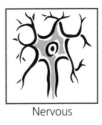

Digestive

Nervous

Endocrine

The genito-urinary system is a combination of the reproductive and urinary systems as a result of the shared position and functions of their joint organs. The reproductive system is responsible for the production of male and female sex hormones and the reproduction of human life. The urinary system is responsible for maintaining balanced fluid levels within the body. This combined system relies on the interaction between the other systems of the body for its survival.

Holistic harmony

The genito-urinary system consists of two closely-linked systems which rely heavily on a balanced network between the other systems of the body to maintain their functions. As a team they share organs and as separate systems they contribute to the healthy development of the human race. Their needs are great and should be taken care of if we are to remain fit and healthy for longer.

Fluid

The balance between water into the body in the form of food and drink and water out of the body in the form of expired air, sweat and faeces is controlled by the kidneys. The amount of water lost through breathing and defecation remains fairly constant except when either system is not functioning well at which point excess water may be lost e.g. stressful situations when breathing levels change or in the case of excessive diahorrea. The amount of water lost through sweating alters as body temperature changes resulting in the following:

- When the body temperature is raised, less urine is formed because more water is lost from the body through sweat which cools the body down.

- When body temperature lowers, less sweat is produced so any excess water is lost through urine.

This activity is controlled by the nervous system (Chapter 9) and the endocrine system (Chapter 10). The brain is able to 'pick up' on the loss of water in the body and alerts the endocrine system, which in turn sends an anti-diuretic hormone, **ADH**, to the kidneys to instruct them to decrease the production of urine. At the same time the brain will activate the thirst sensation. When the fluid levels in the blood are balanced, the production of ADH stops and the action of the kidneys returns to normal.

The body will dehydrate very quickly without water and needs a constant supply to ensure that all body systems and body functions can occur efficiently in order to sustain life. This is even more important during pregnancy as it is thought that the feelings of 'morning sickness' are associated with dehydration.

Nutrition

Water is often referred to as the 'forgotten nutrient' yet it is the most important nutrient after air and contributes to the

Angel advice

Remember that alcohol is a diuretic which has the opposite action of the anti-diuretic hormone. Diuretics stimulate the production of urine. If you have had an evening out on the town consuming large amounts of alcohol, you will most certainly feel the need to urinate throughout the night as well as experience the sensation of extreme thirst! A glass or two of water before going to bed will help balance the situation!

Angel advice

Remember that it is important to increase fluid intake when exercising to avoid dehydration through sweating – the kidneys can only reabsorb water when there is enough water in the system for this to take place!

welfare of the whole body all the time! It is useful to be aware of the following nutritional facts:

- Eating carbohydrates helps to store water in the muscles as glycogen is bound in water.
- Fruit and vegetables contain a large percentage of water. Eating the recommended amount of five servings of fruit and vegetables a day ensures a certain amount of water is contained within the diet.
- Natural spring water provides a certain amount of minerals adding to its nutritional benefits over tap water, which often contains significantly less minerals and also contains other chemicals that are not needed or indeed wanted by the body.
- Fewer minerals are obtained from fizzy water than still water because of the chemical change that takes place when the water is artificially carbonated.

A balanced diet is a requirement throughout life and considerations should be made for the changes that take place in the body as a result of the hormone production in the testes and ovaries. Changes in our diet should accompany changes in hormone levels e.g. more nutrients are required during pregnancy.

Rest

Rest is needed to balance levels of activity within the body. Hormonal secretions in the testes and ovaries will result in changing energy levels and an awareness of this can help to counteract the ill effects by responding to the body's requests for rest. The production of urine is reduced during sleep enabling the body to utilise the water over a period of time. However, if we have drunk more fluid than is needed, the urge to urinate will waken us from even the deepest of sleep! The body will also be forced awake if the opposite occurs and it experiences a dip in fluid levels. The body will alert us to the need to drink by the feeling of extreme thirst.

Activity

Activity helps to keep the body stimulated which in turn affects every system of the body and helps to ensure that specific functions are carried out effectively and efficiently. This is of prime importance in maintaining the reproductive function of humans and the sexual activity between males and female ensures this! It also affects the urinary organs as the production of urine is reduced during

Tip

The kidneys are delicate organs and although they are well protected, care should be taken to avoid applying excessive pressure to them during massage.

Angel advice

To determine the strength of your external sphincter muscle, try the mid-stream test. As you urinate first thing in the morning when your bladder is at its fullest, try to stop mid-flow for a few seconds. If this is difficult to do, then try to ensure that you strengthen the muscle by doing some simple exercises. Squeeze the muscle tightly, hold for a few seconds and release. This can be done at any time and will gradually strengthen the muscle so that the mid-stream test is easy!

exercise to allow the body to lose excess fluid through sweating. Sweating helps to maintain body temperature by cooling the body when we get hot through increased activity. This in turn alerts the kidneys to reduce the amount of water lost in urine in order to ensure that the body's fluid balance is maintained.

Massage is an activity that stimulates the flow of waste around the body by stimulating the circulation. This in turn boosts the activity within the kidneys, increasing the production of urine to rid the body of the unwanted waste. It is for this reason that we often feel the need to pass urine after a massage. It is therefore also necessary to encourage the client to drink plenty of water post-treatment to avoid dehydration and to help 'flush' out the waste from the system.

Air

The heat of air differs from country to country and climate to climate as seasons change. The body gets used to a certain air temperature and the kidneys function accordingly to ensure water and mineral balance is maintained at a constant level. If, however, a person moves to a hot country from a cold country, then the amount of water and minerals lost from the body through sweating increases significantly. An excess loss of sweat leads to an excess loss of minerals, which needs to be balanced out by the kidneys. This re-balancing action occurs over a seven-to-ten-day period as a person acclimatises to their new environment. The body is naturally very adaptable, but it needs time to cope with excessive changes and this can be aided by taking note of what the body needs; we can do this by drinking when we are thirsty, sleeping when we are tired etc.

Age

As with all systems of the body, the genito-urinary system is affected by age and use. Reproductive functions in a female cease with menopause and changes start to take place in the body as the amount of hormones released by the ovaries alters. Men may also experience what is commonly called the 'male menopause' and both sexes experience physical and psychological effects of ageing as a result which are often difficult to come to terms with.

The urinary organs are also affected by age. In a baby the ability to stop urination is not under control and as the bladder fills, a reflex action relaxes the external sphincter in the urethra forcing the expulsion of urine from the body. As an infant develops, it gains voluntary control of the external sphincter muscle, which continues throughout life although age and use can have a gradual debilitating effect on this muscle. For example, pregnancy can put additional strain on the muscle as the baby presses on the bladder; the force of giving birth can relax the muscle; and ageing can greatly reduce the tightness of the muscular action causing periods of incontinence. Exercises can strengthen the muscle and should be done as a matter of course to avoid such problems.

Colour

Red and orange are the colours generally associated with the genito-urinary system. Red is associated with the first chakra and orange with the second chakra. As the organs of the genito-urinary system are distributed among the position of these chakras, the associated colours are thought to have an effect on sexuality and reproduction as well as aiding elimination of excess water and waste from the body. The first chakra is also known as the base or root, which links with the creation of new life. Pink, which is a derivative of red, creates a softer effect and can be used to encourage the nurturing feelings associated with parenting; this can happen through visualisations or it can be worn as an item of clothing. The use of colour can provide a means of enhancing or changing a person's mood and is something that we often do without any conscious thought as our bodies pick up on the various vibrations appropriate to our needs.

Awareness

Children respond naturally to their body's needs and desires without giving it another thought. But as we develop through childhood, our culture forces us to exert a certain amount of control over when and where we perform certain bodily functions. Levels of control are often exceeded and body systems suffer as a result. How often have you found yourself saying that you are too busy to go to the loo? Now that you are aware of the functions of the genito-urinary system, imagine the build up of toxins that will come about if urine is not released regularly. Imagine what that is doing to your insides! How often have you found yourself ignoring the signals of thirst because it is not convenient? Now imagine the concentration of the urine formed as a result of a lack of water in the body. Ignoring these vital signs displays a lack of control in preserving the functions of the body, which is a poor substitute for the gains made in time!

It is equally important to be aware of the changes forced upon the body as a result of changing hormone levels in the testes and ovaries as they have a massive effect on our physical and emotional well-being, especially in women. They initiate the various sexual changes that take place over a lifetime, which can often happen before we feel emotionally ready. Increased awareness leads to increased preparation, and a level of balance, which may make the difference between acceptance and non-acceptance.

Special Care

Good posture contributes to the general well being of the genito-urinary system, as does efficient breathing.

- Good posture allows the organs of the genito-urinary system to sit comfortably within the body. Bad posture forces organs to compress against each other putting added pressure on them.
- Efficient breathing using the diaphragm, intercostals and abdominal muscles gently massages the internal organs of the genito-urinary system and aids their peristaltic action.

The special care of body systems results in the body systems' care of the whole body!

Treatment tracker

GENITO-URINARY SYSTEM

Make up

Acne is often a resulting condition of puberty. The appearance can be helped with the application of corrective make up helping to relieve the stress associated with the condition.

Facials

Facial products such as exfoliators and masks can help the symptoms of acne helping to raise self esteem and confidence levels.

Nail care

Hand and arm and foot and leg massage helps to eliminate the build up of waste in the tight, tense muscles. Drinking a glass of water after such a treatment will help the kidneys 'flush out' the waste.

Hair removal

Hormone imbalance in the ovaries may result in excessive hair growth, e.g. hursuitism in females. Hair removal treatments can help to alleviate the stress associated with such conditions.

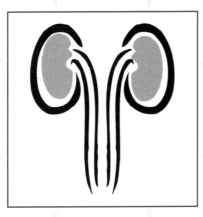

Care should be taken when using electrical treatments such as G5 and vacuum suction over the kidneys to avoid damage caused by excess pressure. Electrical treatments should be avoided over the abdomen during menstruation and pregnancy.

Massage aids good posture which in turn will aid the efficient working of the genito-urinary system by preventing the organs from being restricted.

Massage

Reflexology can help to balance the genito-urinary system improving:

- The removal of waste from the body.
- Hormone production.
- The stages of reproduction.

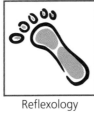

Reflexology

The use of essential oils can help all aspects of the genito-urinary system. As with all systems, oils should be chosen with care especially during pregnancy.

Aromatherapy

Electrical

The genito-urinary system relies on the benefits of relaxation to maintain its many functions. Treatments contribute greatly to the sense of relaxation.

Knowledge review – Genito-urinary system

1 Name the parts that make up the genito-urinary system.

2 Name the male and female genitalia.

3 What are the mammary glands?

4 What do the testes and the ovaries produce?

5 Name the inner concave section of the kidneys and what does it open into?

6 Name the inner and outer layers of the kidneys.

7 What are nephrons and where are they found?

8 Name the tubes that transport urine from the kidneys to the bladder.

9 What is the purpose of the folds in the structure of the bladder and what are they called?

10 Is the urethra longer in a woman or a man?

11 The urethra in a man has a dual purpose – what is it?

12 List the functions of the genito-urinary system.

13 What is a glomerulus and where would it be found?

14 What is the Bowman's capsule?

15 List three substances that are transferred out of the blood in the kidneys.

16 What happens to the glucose, amino acids and vitamins in the kidneys?

17 What happens to urine production when the body sweats excessively?

18 What is the name of the muscle that is under voluntary control enabling the release of urine to be delayed until convenient?

19 What is a good source of fluid in the diet other than water?

20 What effect does alcohol have on the body's fluid balance?

The nervous system

9

Learning objectives

After reading this chapter you should be able to:

- **Recognise the nervous tissue that makes up the nervous system**

- **Identify the parts that make up the central, peripheral and autonomic nervous systems**

- **Understand the functions of the nervous system and the sensory organs**

- **Be aware of the factors that affect the well-being of the nervous system**

- **Appreciate the ways in which the nervous system works with the other systems of the body.**

The next two chapters explore the systems responsible for external and internal communication and control. The first system controls the body with electrical impulses and is known as **THE NERVOUS SYSTEM.** The nervous system allows the body to detect external and internal changes through the sensory organs, i.e. skin, eyes, ears, nose and tongue, and to make use of the information received to form a response by stimulating the muscles and organs into action. It forms vital links with all of the systems of the body providing them with the ability to 'feel' and to make the suitable response.

Science scene

Structure of the nerves, CNS, PNS and ANS

The nervous system can be divided into three main parts:

1. **THE CENTRAL NERVOUS SYSTEM (CNS)** consisting of the brain and spinal cord.

2. **THE PERIPHERAL NERVOUS SYSTEM (PNS)** consisting of nerves which link the various parts of the body with the central nervous system. The **PNS** can be further divided into two parts: **sensory** or **afferent** nerves which take messages from the sensory organs to the **CNS**, and **motor** or **efferent** nerves which relay messages from the **CNS** to the muscles and organs to activate a response.

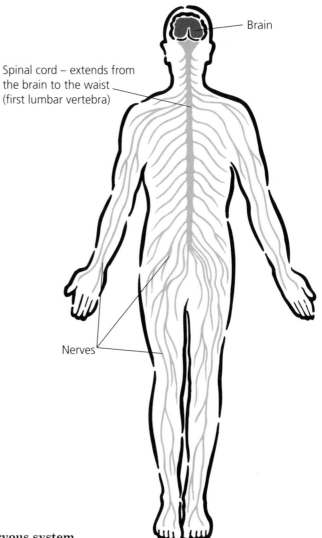

Brain

Spinal cord – extends from the brain to the waist (first lumbar vertebra)

Nerves

The nervous system

3. THE AUTONOMIC NERVOUS SYSTEM (ANS)

consisting of the *sympathetic* and *parasympathetic* systems. These systems work in opposition with one another, the sympathetic nervous system is responsible for preparing the body for action and the parasympathetic nervous system prepares the body for rest.

Together these structures form the complex network that is the nervous system and are responsible for helping to maintain **homeostasis** – a steady state of balance.

The nervous system as a whole is made up of nervous tissue which consists of nerve cells called **neurons**.

Neurons

Neurons are cells which are unique to the nervous system. They are similar to other cells of the body in structure but vary in size, shape and functions. Neurons are long (sometimes up to one metre), narrow in shape and extremely delicate. They are generally unable to renew themselves when destroyed and it is for this reason that

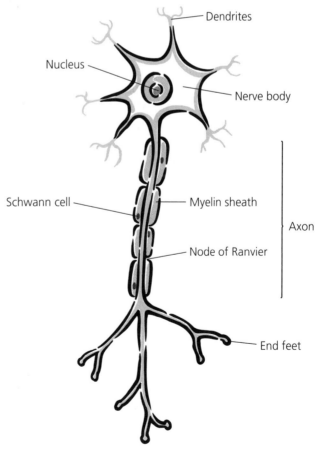

Neuron

disorders of the nervous system resulting in paralysis are often permanent.

Neurons transmit messages to and from the central nervous system (brain and spinal cord) in the form of *impulses*. They are able to 'pick up' on external and internal physical information relating to the body from the sense organs e.g. skin, ears, eyes, tongue and nose; this information is then changed into electrical signals which are passed through the system from neuron to neuron as impulses.

Neurons are made up of a *cell body* containing a large nucleus, as well as bundles of *nerve fibres*. There are two types of nerve fibres:

- **Dendrites** – which pick up impulses and pass them *to* the cell body
- **Axons** – which carry impulses *away* from the cell body.

Each neuron generally consists of many dendrites but only one axon.

A fatty substance called **myelin** forms a white sheath covering the axon of some neurons, insulating them and helping to increase the speed at which the impulses pass through the cell. The myelin sheath is formed in sections along the axon by a **Schwann cell** which wraps itself around the axon. The junction of each section of the myelin sheath is called a **node of Ranvier**. These speed up the passageway even further ensuring that the impulses reach their destination as quickly as possible.

Some neurons do not have a myelin sheath and these are known as *unmyelinated* and their performance is slower.

At the end of the axons there are tiny fibres called *fibrils* also known as end feet. These are responsible for passing the impulses onto the dendrites of the next neuron.

Between each neuron there is a junction called a **synapse**. As an impulse reaches the synapse, a chemical is released called a **neurotransmitter** which enables an impulse from one neuron to be passed to the next by a process known as diffusion (Chapter 1).

Neurons are supported by **neuroglial cells**, a type of connective tissue found exclusively in the nervous system. These cells fill the spaces between neurons, providing a structural framework, and 'mop up' damaged cells and foreign particles as part of phagocytosis.

Groups of neurons form nerves and there are five different types of nerves and nervous tissue forming the nervous system. These include:

Fascinating Fact

Myelin begins to form over nerve fibres during the fourteenth week of foetal development and is not fully completed until a child reaches puberty.

Fascinating Fact

Impulses are sent through the nervous system rather like a relay race with each neuron passing the impulse to the other as quickly as possible. This allows the brain to receive the impulse, e.g. a taste, interpret it and send back an impulse in response in the form of action e.g. spit out the food because it tastes awful etc.

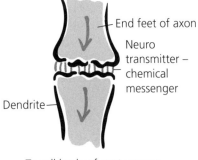

From cell body of one neuron

End feet of axon

Neuro transmitter – chemical messenger

Dendrite

To cell body of next neuron

Synapse

- Sensory or afferent nerves carrying impulses from the body to the CNS i.e. *to the brain.*

- Motor or efferent nerves carrying impulses from the CNS to all parts of the body i.e. *away from the brain.*

- Mixed nerves consisting of both afferent and efferent nerves. These are found in the spinal cord where there is a two-way track enabling impulses to pass both ways.

- White matter – bundles of nerve fibres with a myelin sheath found on the *inside* of the brain and the *outside* of the spinal cord providing a link between the parts of the CNS.

- Grey matter – cell bodies together with dendrites and axons without the myelin sheath. Grey matter is found on the *outside* of the brain and *inside* of the spinal cord and is responsible for co-ordinating the action of the CNS.

The central nervous system (CNS)

The CNS consists of the brain and the spinal cord which are continuous with one another. Both structures are protected on the outside by skin (Chapter 2), muscles (Chapter 4) and bones (Chapter 3).

Beneath this, protecting the brain and spinal cord further, are three layers of tissue collectively known as **meninges.**

Meninges

1. The outer layer or **dura mater** is made up of fibrous tissue containing blood vessels and nerves. It forms the periosteum of the skull bones (Chapter 3) and covers the outer portion of the brain extending down as a strong sheath surrounding the spinal cord.

2. The middle layer or **arachnoid mater** forms a further protective covering over the brain and length of the spinal cord. It forms a **subarachnoid space** which is filled with a fluid similar to blood plasma called **cerebrospinal fluid.** This fluid helps to nourish the CNS, and to support and protect the structures.

3. The inner layer or **pia mater** provides a further protective covering to the brain and spinal cord together with a rich blood supply which supplies vital nourishment.

Activity

Test your knowledge of the skeletal and muscular systems by recalling the bones and muscles of the skull, the five regions of the spine and the muscles of the back. These structures all help to keep the structures of the central nervous system intact.

Fascinating Fact

A blow to the head can cause blood vessels in the meninges to burst. Blood accumulates, affecting the circulation of the cerebrospinal fluid and increasing the pressure between the bones of the skull and the soft tissue of the brain. Unless the blood is removed, the compression effect on the brain can cause severe damage which often results in death. Performing a lumbar puncture, which involves inserting a hollow needle into the subarachnoid space between the 3rd and 4th or 4th and 5th lumbar vertebrae, can check the pressure of the cerebrospinal fluid.

The brain

The brain forms the main part of the central nervous system and is made up of many billions of neurons. It consists of three main sections:

1. The **cerebrum**, also known as the forebrain, makes up the largest part of the brain and is divided into two halves – the left and right **cerebral hemispheres**. The cerebrum is divided up further into lobes which take the names of the corresponding bones (frontal, parietal, temporal and occipital) and are responsible for different activities. These activities are controlled by an inner region of white matter which forms a connection between the two hemispheres called the **corpus callosum**, together with an outer covering of grey matter known as the **cerebrum cortex** in which can be found the **hypothalamus.** The hypothalamus links the nervous and endocrine systems together (Chapters 9 and 10). The grey matter of the cerebrum is associated with mental activities, sensory perception and movement, while the white matter provides a link between the hemispheres and lobes co-ordinating our thoughts, feelings and actions.

2. The **cerebellum** is also known as the small brain; it lies below the cerebrum and is made up of grey and white matter. It is also divided into the left and right hemispheres. Collectively it is responsible for ensuring that our body movements are smooth and co-ordinated and for maintaining balance, posture and muscle tone.

3. The **brain stem** is made up of three parts – the **mid brain** which lies in between the cerebrum and

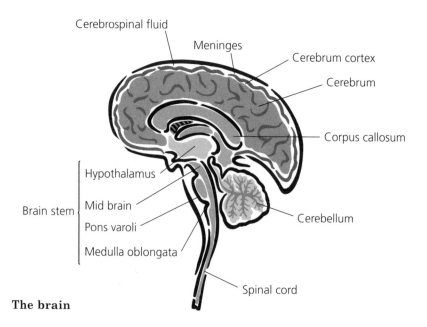

The brain

cerebellum underneath which lies the **pons varolii** and finally the **medulla oblongata.**

- The mid brain transmits impulses linking together the spinal cord, the cerebellum and the cerebrum in both directions.

- The pons varolii forms a bridge between the two hemispheres of the cerebellum transmitting impulses to and from the spinal cord.

- The medulla oblongata consists of white matter on the outside and grey matter on the inside and is continuous with the spinal cord which lies directly below it. It controls the involuntary actions of the respiratory and circulatory systems and some reflex actions e.g. sneezing, coughing, swallowing and vomiting.

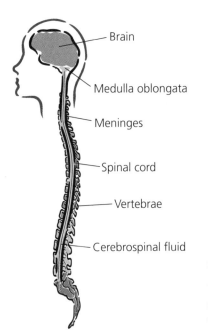

The spinal cord

The spinal cord

The medulla oblongata extends to form the spinal cord which ends at the first lumbar vertebra. It is long and cylindrical, surrounded by the meninges and protected on the outside by the vertebrae bones of the spine. The spinal cord is bathed in cerebrospinal fluid and is made up of many millions of neurons forming grey matter in the shape of a letter H within the middle of its structure with white matter surrounding it. The spinal cord provides the link between the nerves of the body and the brain.

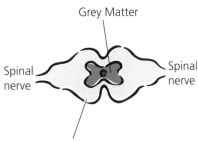

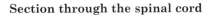

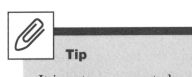

Section through the spinal cord

The peripheral nervous system (PNS)

The PNS contains the nerves of the body which transmit messages in the form of impulses to and from the central nervous system. It consists of twelve pairs of **cranial nerves** which radiate out from the brain to all areas of the face and neck, and thirty-one pairs of **spinal nerves** which radiate out from the spine to the rest of the body. Together, these nerves supply the sense organs and muscles of the body with the ability to 'sense' a stimulus and 'act' upon it by sending messages in the form of impulses to and from the brain.

Cranial nerves

There are twelve pairs of cranial nerves arranged in the following way:

1. *Olfactory nerves* – associated with our sense of smell

2. *Optic nerves* – associated with our sense of sight

Tip

It is not necessary to learn the names of these nerves but a general awareness of their position and action is useful when performing facial massage as our movements stimulate their activity.

3. *Oculomotor nerves* – associated with the movement of the eyes and eyelids

4. *Trochlear nerves* – assist with the movement of the eyes

5. *Trigeminal nerves* – associated with activity in the eyes and the jaw

6. *Abducent nerves* – help with eye movements

7. *Facial nerves* – associated with the taste buds, the salivary glands, the tear ducts and facial expressions

8. *Vestibulocochlear nerves* – help with our sense of balance and hearing

9. *Glossopharyngeal nerves* – associated with the tongue and the pharynx and help to control swallowing

10. *Vagus nerves* – associated with speech and swallowing, as well as being associated with the heart and the smooth muscles of the thorax and abdomen

11. *Accessory nerves* – associated with the neck and back

12. *Hypoglossal nerves* – associated with speaking, chewing and swallowing.

Spinal nerves

There are thirty-one pairs of spinal nerves arranged in the following way:

- 8 pairs of cervical nerves
- 12 pairs of thoracic nerves
- 5 pairs of lumbar nerves
- 5 pairs of sacral nerves
- 1 pair of coccygeal nerves

These spinal nerves intermingle to form networks or groups called **plexuses** from which they divide and branch out to provide nerves to all parts of the body. The nerves of each area of the spine form plexuses except the nerves of the thoracic region. The thoracic nerves provide individual nerve supply to the muscles of the chest and abdomen.

Plexuses

The main plexuses include:

- The *cervical plexus* – containing the first four cervical nerves which branch out to supply the skin and muscles of the neck and shoulder. The *phrenic*

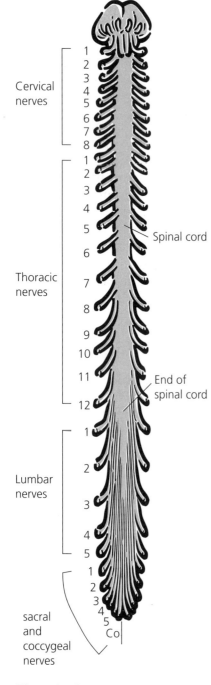

Cervical nerves

1
2
3
4
5
6
7
8

Thoracic nerves

1
2
3
4
5   Spinal cord
6
7
8
9
10
11   End of spinal cord
12

Lumbar nerves

1
2
3
4
5

sacral and coccygeal nerves

1
2
3
4
5
Co

The spinal nerves

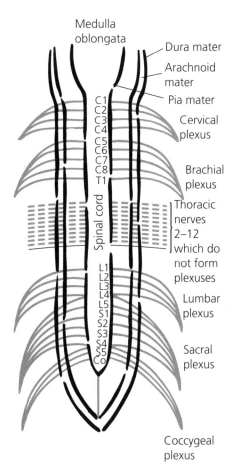

Medulla oblongata

Dura mater

Arachnoid mater

Pia mater

C1
C2
C3
C4
C5
C6
C7
C8
T1

Cervical plexus

Brachial plexus

Spinal cord

Thoracic nerves 2–12 which do not form plexuses

L1
L2
L3
L4
L5
S1
S2
S3
S4
S5
Co

Lumbar plexus

Sacral plexus

Coccygeal plexus

The spinal plexuses

Fascinating Fact

The sciatic nerve is the longest nerve of the body and is responsible for the pain associated with the disorder sciatica. The nerve supply becomes impeded by excess pressure or damage and pain is experienced in the hips and buttocks down through the knee to the ankle.

nerve is associated with this plexus and it stimulates the contraction of the diaphragm during respiration.

- The *brachial plexus* – containing the remaining four cervical nerves and part of the first thoracic nerve which branch out to supply the skin and muscles of the upper limbs. Nerves from this plexus include the radial and ulnar nerves that affect the triceps and biceps as well as the skin and muscles down the arm to the fingertips.

- The *lumbar plexus* – containing the first three lumbar nerves and part of the fourth which branch out to supply skin and muscles of the lower abdomen, groin and part of the lower limbs.

- The *sacral plexus* – containing part of the fourth lumbar nerve, the fifth lumbar nerve and the first four sacral nerves which branch out to supply the skin and muscles of the pelvis, the buttocks and part of the lower limbs. The sciatic nerve is one of the nerves that branches out from this plexus.

- The *coccygeal plexus* – containing part of the fourth and fifth sacral nerves and the coccygeal nerves which branch out to supply the muscles and skin of the external structures of the digestive and reproductive organs.

The peripheral nervous system contains both sensory (afferent) nerves which allow impulses to be sent from the body to the spine and brain and motor (efferent) nerves which allow impulses to be sent from the brain and spinal cord to the various parts of the body.

The autonomic nervous system (ANS)

The autonomic nervous system supplies nerves to all of the internal organs of the body that are not under our conscious control; this means that involuntary actions such as peristalsis are controlled by the ANS. The autonomic nervous system is controlled by the hypothalamus in the cerebrum and is divided into two opposing parts – the sympathetic and parasympathetic systems. Both parts supply nerves to every organ.

Sympathetic nervous system

The sympathetic nervous system consists of a network of nerves which lie in front of the vertebrae from the thoracic to the lumbar region. They form plexuses which provide a network of nerves branching out to supply the organs of the body.

An important plexus of the autonomic nervous system is the **solar plexus**. It is situated in the abdomen below the diaphragm line and is often referred to as the great sympathetic plexus. Solar refers to the sun and the effects of this plexus radiate out like the rays of the sun stimulating many of the major organs of the body.

Tip

The sympathetic nervous system stimulates the activity in the organs needed to ensure that a person is able to deal head on with a stressful situation (fight) or have the energy to run away (flight)!

Tip

The action of the parasympathetic nervous system is to prepare the body for rest and is aided by relaxing treatments such as massage. Massage helps to counteract the stresses of everyday life by encouraging a client to allow the parasympathetic nervous system to be activated.

The hypothalamus uses its links with the endocrine system to initiate the release of the hormone **adrenalin** from the **adrenal glands** (Chapter 10); this stimulates the plexuses of nerves responsible for activating the body to deal with stressful conditions by:

- Increasing heart rate and blood pressure, thereby diverting blood flow to the heart and skeletal muscles and away from the skin and digestive organs
- Increasing intake of oxygen and output of carbon dioxide by dilating the bronchi making it easier to breathe in and out
- Speeding up the production of energy through the conversion of glycogen in the liver
- Delaying digestion as blood is diverted to other organs
- Increasing muscle tone in the urethral and anal sphincters inhibiting micturition and defecation
- Dilating the pupils and opening the eyes wide to increase vision
- Increasing sweating
- Contracting the arrector pili muscles in the skin to form goose pimples.

Parasympathetic nervous system

The parasympathetic nervous system consists of a corresponding network of nerves and is responsible for providing the opposite reaction to that which is produced by the sympathetic nervous system. Following a stressful situation, the hypothalamus stops the release of adrenaline from the adrenal glands and the parasympathetic system takes over. It is known as the 'peace maker' as it calms the body down, counteracting the stimulating effects of the sympathetic system and allowing the body to relax by:

- Reducing heart rate and blood pressure
- Slowing down breathing as less intake and output of air is required
- Increasing digestion and absorption of food as less blood is needed by the heart and muscles
- Relaxing the control over the release of urine and faeces from the urethral and anal sphincters
- Contracting the pupils and relaxing the eyelids giving the appearance of sleepiness.

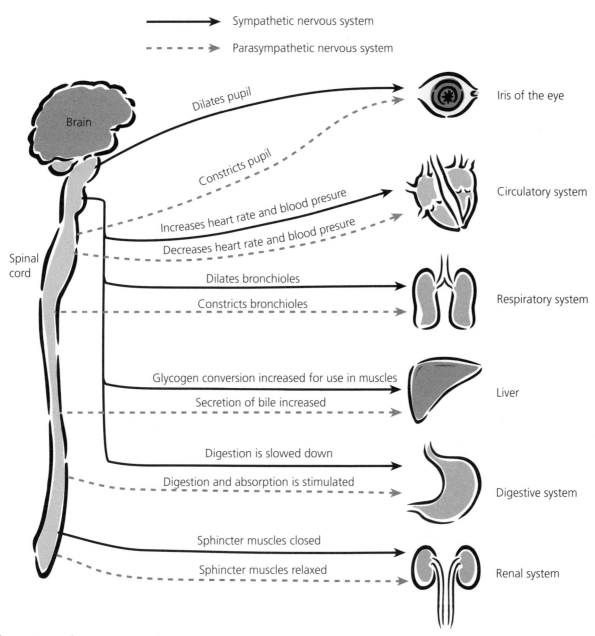

The autonomic nervous system

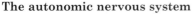

Functions of the nervous system

There are five main functions of the nervous system: sensory, integration, motor, reflex and regulation.

Sensory

Sensory neurons are located within the special sense organs e.g. ears. The ends of the dendrites form *sensory receptors* which 'pick up' on the changes affecting each sense organ e.g. sound. This information is sent as an impulse along the

The neurons and the surrounding tissue fluid both contain positively charged ions. Neurons contain potassium ions and the tissue fluid contains sodium ions. When changes take place e.g. in temperature and pressure, the sodium ions enter the neurons leaving the tissue fluid negatively charged as a result. This change in polarity creates an electrical reaction in the form of an impulse and continues along the length of the neurone and from neurone to neurone, with the help of neurotransmitters.

The limbic centre of cerebrum is also associated with emotion and memory so this enables us to link a specific odour with an emotion!

The structure of the nose is detailed in Chapter 5.

dendrites to the cell body where it is sent along the axon to the end feet, where a chemical neurotransmitter passes it to the dendrites of the next neurone. This process continues along the peripheral nerves to the spinal cord and eventually reaches the brain.

Sense organs

These include the nose, tongue, eyes, ear and skin.

Nose

Olfactory refers to the sense of smell and is achieved by the olfactory organs in the nose:

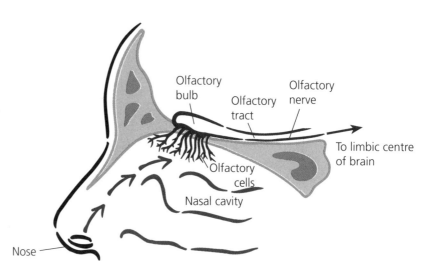

The nose

- Chemicals that stimulate the sense of smell enter the nose as gases in the air.
- The mucous lining of the nose moistens the air dissolving the gases into tiny chemical particles.
- The cilia in the nose are extensions of nerve fibres which are able to detect the different odours associated with each chemical particle.
- Special **olfactory cells** at the back of the nose carry on the action of the cilia by sending the information of the incoming odour to the **olfactory bulb** in the brain to be analysed.
- The information is then sent along **olfactory tracts** via the **olfactory nerves** (1st cranial nerves) to the **limbic centre** in the cerebrum or forebrain where interpretation of the odour takes place.

Fascinating Fact

Endorphins are chemical substances (neuro transmitters) produced at the synapses which raise the pain threshold. They are produced when people are said to have 'reached the pain barrier' and allow a person to continue with their actions without experiencing further pain.

Tongue

The upper surface of the tongue is covered in tiny projections known as papillae (Chapter 7). Within these papillae lie tiny **taste buds**. They are round in structure and form bundles of cell bodies and nerve endings of the 7th, 9th and 10th cranial nerves. The cells contain taste hairs which project upwards towards tiny pores on the tongue's surface. The taste hairs are stimulated by the food and drink we place into the mouth sending messages in the form of electrical impulses to the taste area in the cerebrum for interpretation. Different areas of the tongue respond to different tastes:

- Sweetness is detected by taste buds located at the tip of the tongue.
- Sourness and saltiness is detected by taste buds at the sides of the tongue.
- Bitterness is detected by taste buds at the back of the tongue.

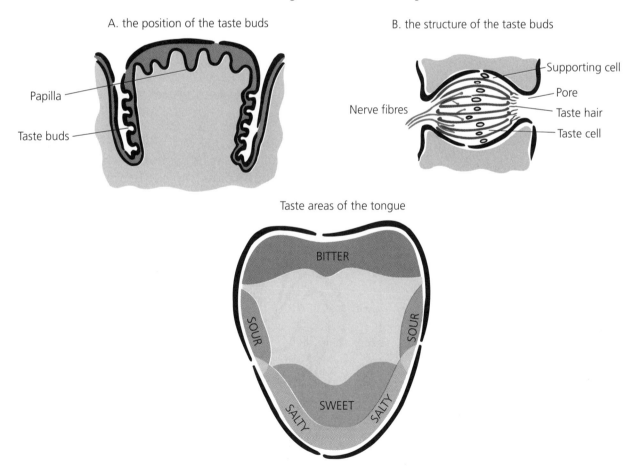

Section through the surface of the tongue to show:

A. the position of the taste buds

Papilla
Taste buds

B. the structure of the taste buds

Supporting cell
Pore
Nerve fibres
Taste hair
Taste cell

Taste areas of the tongue

BITTER
SOUR
SOUR
SALTY
SWEET
SALTY

The tongue

The eye

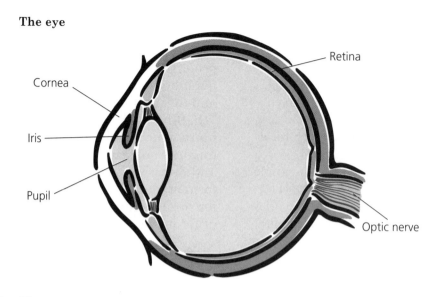

Eyes

The eyes are contained within sockets created by the bones of the skull (Chapter 3). Each eye is shaped like a small ball and contains the **cornea**, **iris**, **pupil** and **retina**. The optic nerves (2nd cranial nerves) connect the eye with the brain. The eyes receive light rays through the transparent cornea. The coloured part of the eye is known as the iris and this regulates the amount of light entering the eye by altering the size of the pupil. The retina is the inner most layer of the eye and contains light-sensitive cells which convert the incoming light rays into nerve impulses and send them to the cerebrum via the optic nerve to interpret what is being seen.

Ears

The external portion of the ear or earflap is known as the **pinna,** the **auditory canal** and the ear drum. The

The ear

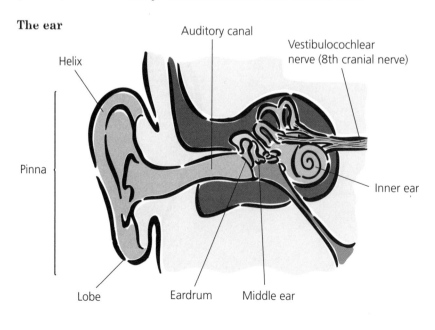

internal portion of the ear is made up of the middle and inner ear.

The pinna consists of a lower section or earlobe and an upper section or **helix**. The earlobe is composed of fibrous and adipose tissue with a rich blood supply and the helix is composed of elastic cartilage with a poor blood supply.

The auditory canal forms an s-shaped tube leading into the ear drum and the middle and inner ear from the pinna.

The ears are associated with the sensory functions of balance and hearing:

1. Balance – the ears detect changes in the position of the head and send messages via the 8th cranial nerve to the cerebellum and cerebrum. The messages are interpreted and the skeletal muscles are instructed to maintain posture and thus balance. Loss of balance occurs when the brain cannot keep up with the changes in movement of the head e.g. spinning around in a circle – this causes us to lose our balance and fall over!

2. Hearing – sound waves are picked up in the ear and transmitted to the cerebrum via the 8th cranial nerve for interpretation.

Skin

Sensory nerve endings in the skin are sensitive to touch, pain and changes in temperature (Chapter 2).

Integration

The brain receives the various impulses via the sensory nerves from the sense organs. The impulses are integrated, translated and stored. As a result, a conscious or unconscious course of action is formulated in the form of responding impulses. The brain gets used to constant or frequent stimuli over time and eventually sensory adaptation takes place. This means that the effect of the stimulation decreases e.g. we soon get used to the feel of hands massaging the body, the smell of perfume etc.

Motor

The responding impulses leave the central nervous system via motor nerves which follow the direction of the peripheral nerves back to the muscles and organs. The impulses pass from one neurone to another with the help of neurotransmitters until they reach the **effector** – a muscle or an organ that will be stimulated to perform the action instructed by the impulses.

Some of these actions are under voluntary control e.g. walking down the stairs.

Other actions involve the autonomic nervous system in order that their performance is involuntary without any conscious thought or effort e.g. the movement of nutrients through the digestive system.

Reflex

The nervous system is able to respond at great speed to an external or internal stimulus in the form of a reflex action e.g. if you reach to pick up a hot dish that has come directly from the oven, your hand will automatically pull away as soon as you detect the heat. The nervous system is able to create a simple pathway called a **reflex arc**, whereby the receptor at the end of a sensory nerve in the skin of the hand picks up on the stimulus (the hot dish) sending the impulses along the neurons to the spinal cord. At this point a reflex centre allows the impulse to by-pass the brain as it sends a responding impulse along a motor nerve to the effector activating an automatic withdrawal response. Reflexes are functions used to describe the subconscious responses of the autonomic nervous system but also incorporate the automatic actions of swallowing, vomiting, coughing and sneezing as well as the knee-jerk reflex.

Reflexes enable the body to limit the damage caused by different stimuli as well as allow the body to perform various functions without conscious effort.

Regulation

The nervous system uses all its parts to regulate the systems of the body to achieve homeostasis in the following ways:

- The CNS regulates the activity within the nervous system as a whole e.g. the hypothalamus in the brain has control of the ANS.
- The PNS regulates the sensory and motor activity of the body e.g. the sense organs respond to a stimulus by sending impulses to the brain via sensory nerves and a response is relayed back via motor nerves.
- The ANS regulates the involuntary activity within the body e.g. breathing, digestion etc.

Common conditions

An A–Z of common conditions affecting the nervous system

- ALZHEIMER'S DISEASE – gradual shrinking of the brain where the nerve fibres become tangled resulting in a progressive decline in mental activities.
- BELL'S PALSY – inflammation of the facial nerve resulting in sudden paralysis causing one side of the face to drop. Complete recovery usually occurs within a few weeks.
- CATAPLEXY – a sudden collapse of posture as a result of an extreme emotion e.g. sadness, anger, excitement.
- CEREBRAL PALSY – a permanent disorder of the brain affecting the control of muscles. Muscles have reduced control and go into spasms.
- CLUSTER HEADACHE – severe headache that starts three to four hours after falling asleep, recurs nightly for weeks or months and then disappears for years. More common in men than in women.
- DELIRIUM TREMENS (DT's) – confusion, hallucinations and trembling associated with the withdrawal symptoms which occur when an alcoholic stops drinking.
- DEMENTIA – gradual death of brain cells through normal ageing. Memory loss, confusion and changes in behaviour may result.
- EPILEPSY – temporary loss of consciousness. It is known as *petit mal* when loss of consciousness is for a few seconds and *grand mal* when unconsciousness is accompanied by convulsions.
- EXTRADURAL HAEMATOMA – a complication of an injury to the head whereby one of the bones of the skull is fractured rupturing blood vessels and producing a blood clot which exerts pressure on the brain.
- FALLS – old people may suddenly fall as a result of passing deficiencies in the cerebral circulation.
- MENINGITIS – a severe infection of the membrane surrounding the brain and spinal cord.
- MIGRAINE – recurrent severe headache with additional symptoms including flashing lights before the eyes together with a dislike of bright lights. Nausea and vomiting may also occur.
- MOTOR NEURONE DISEASE – a condition which affects the motor neurons causing progressive weakness in the muscles.

- MULTIPLE SCLEROSIS (MS) – degeneration of the nervous tissue in the central nervous tissue. This disease starts in adults between the ages of 20–50 and affects the parts of the body controlled by those sections of damaged nervous tissue including vision, speech, movement etc.

- MYALGIC ENCEPHALOMELITIS (ME) – also referred to as *post-viral fatigue* and resembles the symptoms that follow many viral infections including muscle pain, fatigue, exhaustion, depression etc. Symptoms may last for several months and even years.

- NEURALGIA – pressure on a nerve caused by irritation. Pain may be felt along the length of the nerve as well as at the point of pressure.

- NEURITIS – inflammation of nerves resulting in muscle weakness and loss of skin sensitivity.

- NEUROSIS – excessive feelings of anxiety, depression and/or phobia.

- NOCTURNAL MYOCLONUS – a sudden jerking of muscles as a person is drifting off to sleep causing momentary panic. If this occurs frequently it can interfere with sleep.

- PARKINSON'S DISEASE – a condition in which muscular stiffness and tremors develop as parts of the base of the brain degenerate leading to a deficiency of dopamine which aids the transmission of nerve impulses.

- SCIATICA – abnormal pressure on any part of the length of the sciatic nerve, which extends down the legs from the lower back, resulting in pain.

- SPINA BIFIDA – a condition that is present at birth. Spinal nerves are affected due to malformation of bones and tissue surrounding the spine resulting in mental and/or physical handicap.

- STROKE – a sudden loss of function on one side of the body caused by an interruption of the blood supply to part of the brain.

- SUBARACHNOID HAEMORRHAGE – rupturing of the blood vessels on the surface of the brain causing bleeding in the space around the brain. Occurs mainly in young adults for no apparent reason.

- TENSION HEADACHE – pain resulting from overworked muscles of the scalp, face and neck often associated with excessive concentration.

- TIC – a nervous habit resulting in a persistent twitch of a muscle.

- VERTIGO – feeling of dizziness when standing still.

System sorter

THE NERVOUS SYSTEM

Skeletal

Muscular
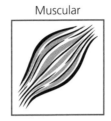

Integumentary

Respiratory
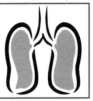

The bones of the skull and the vertebrae of the spine protect the delicate structures of the central nervous system ie brain and spinal cord.

The brain sends messages as impulses to the muscles via efferent motor nerves which in turn activate movement.

The skin is a sensory organ receiving information from the outside world in the form of changes in temperature, touch, pressure etc which are sent to the brain for interpretation via afferent sensory nerves.

The O_2 processed from the air inhaled into the body is vital for the well-being of all nerve cells that make up the nervous system. Without adequate O_2 they will become damaged and die causing irreversible brain damage.

The hypothalamus in the brain links the nervous and endocrine systems as both systems have a controlling effect on the body as a whole. The nervous system controls by using electrical messages in the form of impulses; the endocrine system controls with the release of chemical messages – hormones.

Blood transports O_2 to the nerve cells. The medulla oblongata and the hypothalamus control the actions of the circulatory systems.

The nervous system influences the actions of the digestive system as the sympathetic NS prepares the body for activity slowing down digestion. The parasympathetic NS prepares the body for rest speeding up digestion.

As the bladder fills with urine the nerve endings alert the brain. A message is relayed to the external sphincter muscle to relax to allow urination to occur.

Circulatory

Digestive

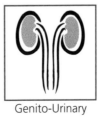

Genito-Urinary

Endocrine

The nervous system is the body's communication system. It is able to inform the brain of any changes happening within and outside of the body, decide what it wants the body to do in response and then activate the relevant muscles and organs accordingly.

Holistic harmony

Although the nervous system is responsible for the control of the whole body, it is itself a very delicate system and one which needs a lot of care.

Fluid

Alcohol and caffeine have a debilitating effect on the nervous system. When they are taken together as part of a meal, the effect is multiplied. This combination contributes to reducing a person's reaction time and can increase the feelings of drunkenness and resulting hangover.

The initial effects of caffeine and alcohol are stimulating; they make a person feel wide awake. However, as caffeine and alcohol also stimulate the production of urine, the body can become dehydrated and the result is often experienced as a headache. The more caffeine and/or alcohol consumed the worse the headache! Drinking water will help to rehydrate the body and ease the headache.

Nutrition

Food plays an important role in maintaining the well-being of the nervous system. Toxins damage nervous tissue and this affects all aspects of the system including intelligence, memory and concentration. A diet high in sugar and the type of refined carbohydrates found in fast foods have a negative effect on mental activity.

The vitamin B group of nutrients is particularly beneficial for brain activity; this includes Vitamins B1, B3, B5, B6 and B12. Good sources of these vitamins include:

- Vitamins B1, B3 and B6 – watercress, cauliflower and cabbage
- Vitamins B1, B3 and B5 – mushrooms
- Vitamin B12 – oily fish, dairy products and poultry.

It is important to note that the benefits of all of these nutrients are destroyed in the body by alcohol and smoking.

Rest

Sleep is necessary for the nervous system as this is the time when the brain sorts out and stores the information it has received throughout the day. The nervous system gets

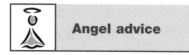

Remember

Remember that alcohol and caffeine are readily absorbed into the blood stream. They are transported to the brain where they are absorbed and as a result, they affect the functions of the nervous system very quickly.

Angel advice

For every cup of coffee and/or alcoholic drink consumed, ensure that a glass of water is also drunk to help balance the effects.

Tip

Think about the amount of concentration it has taken to understand the contents of this book. If you read a chapter at a time and allow your brain to digest the information, you will retain that knowledge far more easily than if you tire the brain by trying to learn the whole book in one session!

Fascinating Fact

More and more businesses are being forced to recognise the need to reduce the stress levels of their workforce and are introducing beauty and holistic treatments as part of a weekly stress relief package.

Angel advice

To exercise the central nervous system: lie on the floor resting your back on a soft mat for support. Bring your knees to your chest and gently cradle your legs with your arms. Rock your body gently from side to side and back and forth. This will help to clear the neuromuscular pathways by easing out tension in the nerves and surrounding muscles, allowing the impulses to pass more freely to and from the brain.

Fascinating Fact

Approximately 2,000 litres of blood pass through the brain every day allowing the absorption of about 60 litres of oxygen.

tired in much the same way as all of the other systems of the body and it needs adequate resting time in order to cope with the pressures it has to endure on a daily basis. The nervous system also benefits from short periods of rest which allow time for relaxation between mental activities. Taking our mind off a task for a while will allow the brain to adjust. This may take the form of relaxing with a magazine or better still practising a few moments meditation.

Resting helps to clear the mind and make space for further input. Treatments such as Indian Head Massage create a resting state within the body and mind, and encourage the parasympathetic nervous system to take over; they can be carried out at any time during the day in order to ease the pressures of a working day.

Activity – the nervous system relies on mental and physical activity to keep it healthy. Boredom results in lethargy and a general lack of interest in life. Keeping active both mentally and physically keeps life interesting and exciting.

Air – the nervous system is reliant on a good supply of oxygen without which nerve cells soon die. As nerve cells cannot generally be replaced, the need for oxygen is vital.

The quality of air breathed into the body is an important factor and care should be taken to avoid pollution and smoking both of which contribute to reducing levels of intelligence, memory and concentration. Taking time to practise the art of breathing will help to clear both body and mind.

Age

As we age there is a tendency toward a reduction in mental function. Reactions often become slower and less co-ordinated as the sensory organs also lose some of their functions. Our sense of sight, hearing, smell and taste suffer with the passing of time and an ageing body experiences varying degrees of difficulty including:

- Difficulty in focusing the eyes on close objects
- A gradual deterioration of hearing
- An inability to detect smells such as gas, body odour, rotting food etc.
- A decreasing sense of taste at the same time as the sense of smell disappears, because of the link between the two.

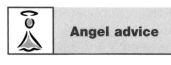

Memory may also be affected as short-term memories become more difficult to retrieve than long-term memories.

As with most other parts of the body, the nervous system relies on use to keep it fit and healthy. The saying 'what you don't use, you lose' is true of this system and should remind us to make constant use of the faculties available to us. Not only will this mean that they will stay in better condition but also that they will be able to perform their functions for longer!

Colour

The colours violet, indigo and yellow are associated with the nervous system generally. Violet corresponds with the seventh chakra which is also known as the crown chakra signifying the position of the brain. Indigo corresponds with the sixth chakra and is strongly linked with the special senses of sight, smell, hearing, taste and equilibrium. Yellow corresponds with the third chakra which is also known as the solar plexus chakra and is strongly linked with the autonomic nervous system. We can add colour to our thoughts by using our various senses e.g. sight and touch. We may also experience the 'thought' of colour as we close our eyes and look into our 'mind's eye'. This facility is often enhanced by the giving and receiving of beauty and holistic treatments. Clients will often say that they 'saw' colours whilst having relaxing treatments such as an Indian head massage, facials, reflexology etc. As a therapist, you may sometimes be able to close your own eyes during certain parts of the treatment in order to change your level of concentration and it is at this point that you may also 'see' colours. This vision of colour often relates to the corresponding area of the body that may be in need of extra treatment or it may provide a means of inter-communication between the therapist and the client, allowing the therapist to 'pick up' intuitively on the way the client is feeling, a bit like picking up on a person's 'vibes', which is in fact exactly what is happening. For some people this skill feels like second nature and one that they are completely comfortable with. For others, however, it seems odd and unnatural. Whatever your thoughts and feelings are, it is wise to have an open mind as many therapists and clients are interested in exploring these skills further, and a general appreciation of the use of colour at many levels is useful even if we chose not to use them ourselves.

Awareness

The nervous system responds to stressful situations by activating the action of the sympathetic nervous system. As a result the body is prepared for the stresses and strains that accompany everyday living. However, many people live their lives under a constant flow of this sympathetic activity which will have a long-term detrimental effect on the body and mind. The body desires balance and this can only be achieved if the compensating action of the parasympathetic nervous system is allowed to function.

It is vital to be aware of the ways in which you can help your body create this level of balance by:

- Avoiding rushing around more than is necessary – this will prevent the muscles from becoming fatigued and help to prevent tension headaches.
- Relaxing when eating to allow the digestive system to work – remember it slows down when the sympathetic nervous system takes over. Spending time over a meal will prevent indigestion and avoid the more serious symptoms associated with irritable bowel syndrome.

These factors contribute to the majority of stress-related problems and can easily be avoided!

Special Care

Care of the nervous system relies on special care of the whole body, as you really cannot have one without the other. There is so much that we do not yet know relating to the functions of the nervous system and the medical profession is continually learning more about the extremely complex activity of the brain. An awful lot goes on within the brain that we do not understand and it is up to us to remain 'open minded' so that we are able to 'take in' some of the things that can sometimes seem beyond the realms of all possibility. As therapists developing our practical skills, we are also developing our mental skills and this includes the development of intuitive thought and action. With the introduction of so many treatments from eastern cultures we are being increasingly encouraged to develop these skills further.

As therapists, we need to develop both sides of the brain equally and take special care to understand the logic behind a new idea or concept whilst at the same time be creative enough to adapt that knowledge to suit both our clients and ourselves.

Fascinating Fact

In the majority of right-handed people, the left hemisphere of the cerebrum of the brain is believed to be connected with logical thinking while the right hemisphere is associated with creativity. We often refer to people as being either more left-brained or more right-brained depending on their levels of common sense!

Treatment tracker

THE NERVOUS SYSTEM

Make up

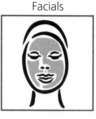

The nervous system is responsible for interpreting the visual image we see when we look in the mirror. The application of make up can help to improve our visual image and boost self esteem and confidence levels.

Facials

A twenty minute massage routine completed during a facial will have a calming effect on the ANS helping the parasympathetic nervous system to slow down breathing, reduce blood pressure and heart rate helping to balance the whole body.

Nail care

Nails can be cut without fear of pain as they do not contain any nerve cells. However, pain would be experienced if nails were pulled out due to sensory nerves in the skin.

Hair removal

The removal of hair is painful due to the sensory nerves present in the skin which detect the pain passing it as impulses to the CNS. Placing a hand down on the skin after removing depilatory wax eases the pain by calming the sensory nerves.

Sensory adaptation occurs during massage as a person becomes accustomed to the feelings of the gentle stroking effleurage movements before experiencing the deeper more stimulating effects of petrissage and tapotement movements.

Massage

Electronic muscle stimulating machines are used to stimulate the motor nerves of a muscle directly. This will cause a contraction in the muscle in the same way as if the brain had initiated the movement.

Electrical

Reflexology stimulates the nerve endings in the feet. This has a clearing effect on the neuromuscular pathways to the CNS.

Reflexology

Essential oils can be used to stimulate and/or relax the nervous system depending on the needs of the client. Oils suitable for stimulating including rosemary and niaouli and for relaxing lavender and camomile.

Aromatherapy

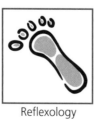

Beauty and holistic treatments contribute to the performance of the nervous system which relies on a balance between relaxation and activity.

Knowledge review – Genito-urinary system

1 Name the three main sections of the nervous system.

2 Name the parts that make up the CNS.

3 Name the two fibres that make up a neurone.

4 What covers some axons and why?

5 What is the action of sensory nerves?

6 What is the action of motor nerves?

7 Name the three layers of the meninges.

8 Where is cerebrospinal fluid found?

9 How many hemispheres and lobes make up the cerebrum?

10 What is white matter and where is it found?

11 What is grey matter responsible for in the cerebrum?

12 Which structure is also known as the 'small brain'?

13 What is the spinal cord an extension of?

14 How many cranial and spinal nerves are there?

15 What is a plexus?

16 Which part of the ANS activates the body in response to a stressful situation?

17 Name the part of the brain that controls the ANS.

18 Which part of the ANS is sometimes referred to as the 'peace maker'?

19 Where is the solar plexus located and which system is it associated with?

20 Name the functions of the nervous system.

The endocrine system

Learning objectives

After reading this chapter you should be able to:

- **Recognise the individual glands and their position within the body**

- **Identify the secretions of each endocrine gland together with their action**

- **Understand the functions of the endocrine system**

- **Be aware of the factors that affect the well-being of the endocrine system**

- **Appreciate the ways in which the endocrine system works with the other systems of the body to maintain homeostasis.**

We are going to carry on our exploration of the controlling and communicating systems of the body by looking at the system responsible for chemical messages. This system is known as **THE ENDOCRINE SYSTEM.** The endocrine system is closely linked to the nervous system (Chapter 9). Together they act as the main communication centres of the body. The nervous system communicates with electrical messengers or impulses via nerves and the endocrine system communicates with chemical messengers or hormones. These chemical messengers are produced by endocrine glands and pass directly into the blood stream to be transported to the part of the body they are to communicate with. The nervous system is involved with the rapid action within the body e.g. movement, and the endocrine system is generally involved with slower adjustments to the body e.g. growth.

Science scene

Structure of the endocrine glands and hormones

The endocrine system consists of a set of endocrine glands which are quite widely spaced from one another and have no direct links. Each gland is responsible for the production of **hormones**.

Fascinating Fact

The term hormone comes from the Greek word *hormaein* which means 'to arouse'.

Hormones

A hormone is a chemical substance which has the ability to affect changes in other cells. They are secreted directly into the blood stream and transported to the various systems of the body. Special cells known as *target cells* receive the hormones and allow the body system to respond to the message and initiate the appropriate changes. Hormones are made from components of the food we eat and are either protein-based (which make up the great majority of all hormones), or fat-based known as steroids.

Tip

Exocrine is the name given to glands that secrete their products into ducts, e.g. sweat glands produce sweat (chapter 2) which exits the body via sweat ducts and bile produced in the gall bladder (Chapter 7) enters the duodenum via ducts.

Endocrine glands

Unlike other glands of the body, endocrine glands are ductless. This means that they secrete their products (hormones) directly into the blood stream.

The main endocrine glands of the body include:

- One pituitary gland located in the brain
- One pineal gland located in the brain
- One thyroid gland located in the neck
- Four parathyroid glands located on the sides of the thyroid gland in the neck
- One thymus gland located in the thorax
- Two adrenal glands located on top of each kidney
- The islets of Langerhans located within the pancreas
- Two ovaries located in the lower abdomen in women
- Two testes located in the groin in men.

The nervous tissue of the hypothalamus in the brain (Chapter 9) creates the link between the nervous and endocrine systems. It forms an attachment with the pituitary gland in the brain allowing two-way communication to take place between the systems. The hypothalamus receives a great deal

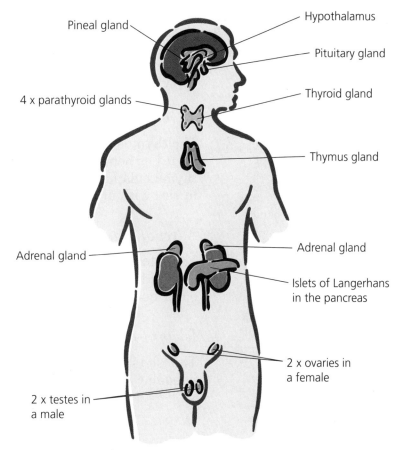

The endocrine glands

of information from the body through its nerve connections in the cerebrum which it then relays to the pituitary gland.

Pituitary gland

The pituitary gland is about the size of a pea. It is situated at the base of the brain behind the nose and is linked to the hypothalamus with which it works as a unit by the pituitary stalk. It consists of two sections or lobes: the anterior (front) and posterior (back) lobes.

Pineal gland

The pineal gland is a tiny gland situated deep in the brain between the cerebral hemispheres. It is also connected to the brain by a short stalk containing nerves which end in the hypothalamus. It is often referred to as the 'third eye' because of its position.

Thyroid gland

The thyroid gland consists of two lobes which take the shape of a butterfly situated just below the larynx and in front of the trachea in the neck.

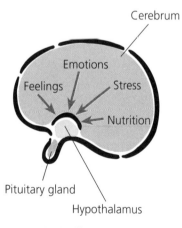

Factors which affect the hypothalamus in turn affect the hormones produced by the pituitary gland

The hypothalamus

Parathyroid glands

There are four parathyroid glands arranged in pairs. Each pair is embedded on either side of the back of the lobes of the thyroid gland.

Thymus gland

The thymus gland is made up of lymphoid tissue (Chapter 6). It lies behind the sternum in between the lungs. The thymus gland is large in children and gradually decreases in size with age.

Adrenal glands

There are two adrenal glands, one situated on the top of each kidney (Chapter 8). Each gland is composed of an outer **cortex** and an inner **medulla**.

Islets of Langerhans

The islets of Langerhans are small clusters of cells found at irregular intervals within the pancreas.

Ovaries

There are two ovaries which form part of the female reproductive system (Chapter 8). They are almond-shaped glands situated within the pelvic girdle on either side of the uterus (womb).

Testes

There are two testes which form part of the male reproductive system (Chapter 8). They are situated in a sac known as the **scrotum** which hangs externally from the body under the penis.

Production of hormones

The hypothalamus starts the process off by producing its own set of hormones as a result of stimulation in the brain. This has a 'knock on' effect on the pituitary gland which in turn produces hormones that stimulate the other glands into producing the relevant hormones to initiate the appropriate changes within the body.

Control of hormone secretion

The secretion of hormones is controlled in three ways. These sources of control include the hypothalamus, the nervous system and the internal environment of the body.

1. The hypothalamus produces hormones of its own called *releasing hormones* which regulate the hormone secretion of the pituitary gland. As a result the pituitary gland then produces hormones which control the other endocrine glands. For this reason, the pituitary gland is often referred to as the **master gland.**

2. The nervous system controls some endocrine glands directly especially when a quick response is required e.g. the sympathetic nervous system stimulates the production of adrenalin in the adrenal medulla in response to a stressful situation (Chapter 9).

3. The internal environment of the body controls the action of some glands e.g. the brain is alerted through the nervous system when changes occur in the levels of sugar in the blood, i.e. a rise or fall. This initiates the production of the hormones in the pancreas which are able to restore the blood sugar levels back to normal.

Hormones are circulated around the body in the blood stream before reaching the target cells of the appropriate organs. Hormones will then pass through the liver where they are broken down and excreted from the body in urine.

Hormone imbalance

Hormone imbalance is created when the endocrine glands produce either too much (over- or hypersecretion) or too little (under- or hyposecretion) of the hormones needed to maintain homeostasis. This results in malfunction and conditions associated with illness and disease.

Endocrine glands and their hormones

Pituitary gland

The anterior lobe of the pituitary gland produces hormones that control the other endocrine glands as well as producing hormones that control other systems. The hormones that control other endocrine glands include:

- **ACTH a**dreno**c**ortico**t**rophic **h**ormone controlling the outer cortex of the adrenal gland
- **TSH t**hyroid **s**timulating **h**ormone – thyrotrophin controlling the thyroid gland
- **Gonadotrophins** controlling the ovaries in women and testes in men.

There are two main gonadotrophins:

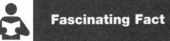

Fascinating Fact

The presence of certain hormones in the body can be determined by testing the urine.

Tip

Gonads are the collective term given to the ovaries in women and the testes in men.

Tip

An imbalance in the secretion of these hormones results in the inefficient production of hormones in the corresponding endocrine glands

1. **FSH f**ollicle **s**timulating **h**ormone which stimulates the ovaries in women to produce the female sex hormone oestrogen and stimulates the testes in men to produce sperm.

2. **LH l**uteinising **h**ormone which stimulates the ovaries to produce the female sex hormone progesterone and the testes to produce the male sex hormone testosterone.

Tip

Imbalance of FSH and LH combined = hypersecretion and hyposecretion can result in infertility; hyposecretion can also result in failure to enter puberty, reduced sex drive, hair loss in males and amenorrhoea in females.

The hormones produced in the anterior pituitary that control other systems include:

- **GH g**rowth **h**ormone – somatotrophin, which promotes the growth of the skeletal and muscular systems (Chapters 3 and 4).

Tip

Imbalance = hypersecretion results in gigantism in children and acromegaly in adults. Hyposecretion results in dwarfism in children and lethargy and weight loss in adults.

- **PRL pr**olactin which promotes the growth of the ovaries, testes and mammary glands and stimulates lactation (milk production) in the breasts as part of the reproductive system (Chapter 8).

Tip

Imbalance = Hypersecretion can result in galactorrhoea, hirsutism, menstrual problems, impotence and infertility. Hyposecretion can result in failure of breast milk production after pregnancy.

- **MSH m**elanocyte **s**timulating **h**ormone which promotes the production of melanin in the epidermis of the skin (Chapter 2).

Remember

Alcohol and caffeine are known as **diuretics** which means that they have the opposite effect from ADH. Diuretics increase the production of urine which can result in the body losing too much fluid and becoming dehydrated.

Tip

Imbalance = hypersecretion results in hyperpigmentation e.g. chloasma and hyposecretion has the opposite effect e.g. vitiligo.

The posterior lobe of the pituitary gland is responsible for the production of two hormones:

- **ADH** **a**nti**d**iuretic **h**ormone – vasopressin which regulates the fluid balance of the body by decreasing urine production in the kidneys (Chapter 8).

Tip

Imbalance = hypersecretion results in oedema (swelling) and hyposecretion results in cranial diabetes insipidus and kidney problems.

- **OT** **o**xy**t**ocin which stimulates the contraction of mammary glands to force the milk from the breast and uterine contraction in the womb in preparation for childbirth. It is also responsible for promoting maternal behaviour.

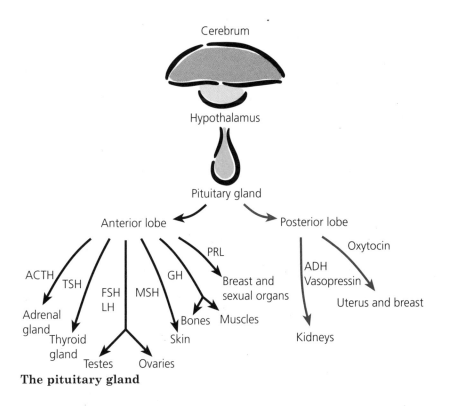

The pituitary gland

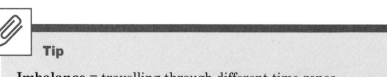

Tip

Imbalance = hypersecretion has no ill-effects and a hyposecretion can cause difficulty in breast-feeding.

Pineal gland

The pineal gland produces the hormone **melatonin** which is often referred to as the chemical expression of darkness. The production of melatonin is stimulated at night when the sun goes down, inducing sleep.

Tip

Imbalance = the pineal gland is thought to be associated with the seasonal affective disorder **SAD**. The darker winter months bring about an increased production of melatonin making a person feel tired and sad!

The pineal gland is responsible for regulating body rhythms including the monthly menstrual cycle as well as the daily sleep/wake cycle.

Tip

Imbalance = travelling through different time zones upsets the pineal gland activity resulting in jet lag.

Thyroid gland

The thyroid gland is responsible for the production of three hormones in response to the production of TSH in the pituitary gland: **thyroxine, triiodothyronine** and **calcitonin.**

- The production of thyroxine and triiodothyronine is stimulated in response to levels of activity within the body in order to regulate metabolism (the rate at which the cells use oxygen).

Activity

When the calcium and phosphorus levels in the body are high, calcitonin stimulates the storage of some of the minerals in the bones and the release of excess minerals in urine.

Tip

Imbalance = hypersecretion produces hyperthyroidism also known as Grave's disease or thyrotoxicosis, resulting in increased metabolism and weight loss. Hyposecretion produces hypothyroidism, known as cretinism in children and myxoedema in adults, and goitre which result in slow metabolism and weight gain.

● The production of calcitonin helps to regulate the calcium and phosphorus levels in the body.

Tip

Imbalance = hypersecretion causes lowering of blood calcium levels and hyposecretion has the opposite effect.

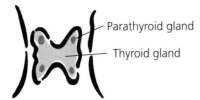

— Parathyroid gland

— Thyroid gland

The thyroid and parathyroid glands

Parathyroid glands

The parathyroid glands produce **PTH parath**ormone which helps to regulate the calcium and phosphorus levels in the blood together with the calcitonin produced by the thyroid gland.

Tip

Imbalance = hypersecretion may cause hypercalcaemia (excess calcium in the blood) resulting in soft bones and spontaneous fractures. Hyposecretion has the opposite effect and produces the condition hypocalcaemia; this causes spasms from over-contraction of the muscles, known as the condition tetany, and over-stimulation of the nerves resulting in convulsions.

Fascinating Fact

When calcium and phosphorus levels in the body are low, PTH will stimulate the reabsorption of these minerals from the bones and decrease the amount lost in urine. PTH also converts vitamin D into a hormone that helps to increase the amount of calcium available.

Thymus gland

The thymus gland is responsible for the production of a group of hormones called **thymosins** which are needed to stimulate the production of **T**-lymphocytes to help in the protection of the body against antigens (Chapter 6).

Tip

The **T** signifies that the lymphocytes are from the thymus gland.

As the thymus gland is larger in children than in adults, it is believed that it has the effect of inhibiting the development of the sex organs until the onset of puberty when it starts to shrink in size.

Adrenal glands

There are three groups of hormones known as steroids produced by the outer adrenal cortex in response to the production of ACTH in the pituitary gland:

- Glucocorticoids, including **cortisol** and **cortisone**, which have an effect on metabolism, development and inflammation.

Tip

Imbalance = hypersecretion can cause Cushing's syndrome. The main features of this condition include obesity, hirsuitism, hypertension, diabetes and osteoporosis. In children it may also cause retarded growth. Hyposecretion can contribute to the causes of Addison's disease.

- **Mineralocorticoids** including **aldosterone,** which help to regulate the concentration of minerals in the body e.g. sodium and potassium.

Tip

Imbalance = hypersecretion causes hyperaldosteronism or Conn's syndrome leading to high blood pressure due to excessive potassium in the blood. Hyposecretion causes Addison's disease leading to weak and wasted muscles and a slowing down of body systems.

- **Sex corticoids**, including **androgens,** which promote the sexual development in males and females including growth of pubic and underarm hair. It is also responsible for sexual drive.

> **Tip**
>
> **Imbalance** = hypersecretion causes hirsutism and amenorrhoea i.e. male characteristics in women, and wasting of muscle bulk and the development of breasts (gynaecomastia) i.e. female characteristics in men. Hyposecretion is contributory to the causes of Addison's disease.

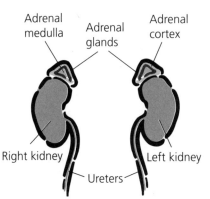

The adrenal glands

The inner adrenal medulla produces two closely related hormones: **adrenalin** and **noradrenalin**. The production of these hormones is stimulated by the sympathetic nervous system in response to stress and helps to prepare the body for 'fight or flight' (Chapter 9).

> **Tip**
>
> **Imbalance** = hypersecretion results in hypertension, headaches, sweating, tremor, nausea, pallor, vomiting and weight loss. The sympathetic nervous system compensates for any deficiencies, with no side effects.

Islets of Langerhans

The islets of Langerhans in the pancreas produce two hormones responsible for regulating the levels of sugar in the blood. These include:

- **Insulin** which is released into the blood stream when there is a rise in blood sugar levels. It promotes the storage of sugar in the form of glycogen in the liver and the muscles, thus reducing the levels in the blood and restoring the balance to normal.

> **Tip**
>
> **Imbalance** = hypersecretion can cause hypoglycaemia (low blood sugar levels) and hyposecretion can cause diabetes mellitus and hyperglycaemia (excess sugar in the blood).

- **Glucagon,** which has the opposite effect on the body from insulin. Glucagon is released into the blood stream when blood sugar levels are low. It promotes the release of glycogen from the liver and muscles increasing the sugar levels in the blood, restoring the balance to normal.

Remember

Females have small amounts of male sex hormones present in the body in the same way as males have small amounts of female sex hormones. This is commonly referred to as the 'masculine' side of women and the 'feminine' side of men.

Fascinating Fact

Anabolic steroids are artificial hormones which replicate the male hormone testosterone. They increase muscle bulk and power but can have severe side effects, including impotence in males and irregular menstrual cycle and the development of male physical characteristics in females e.g. hair growth.

Tip

The menopause is accompanied by reducing levels of oestrogen and progesterone which contributes to the symptoms of hot flushes, sweats, palpitations etc.

Tip

Imbalance = hypersecretion results in hyperglycaemia and any deficiency is compensated for by other hormones e.g. cortisol, GH and the sympathetic nervous system.

Ovaries

The ovaries are responsible for the production of the sex hormones **oestrogen** and **progesterone** in females.

Testes

The testes are responsible for the production of the sex hormone **testosterone** in males.

The sex hormones are responsible for the development of secondary sexual characteristics.

In females these include:

- The development of breasts
- The reproductive organs becoming functional
- The start of the menstrual cycle
- The widening of hips and increase in subcutaneous fat in this area.
- Pubic and axillary (underarm) hair growth.

In males these include:

- The reproductive organs becoming functional
- The production of sperm and semen
- The breaking of the voice
- Increased growth of muscles and bones
- Facial, pubic, axillary, abdominal and chest hair growth.

Tip

Imbalance = hypersecretion of oestrogen in males can lead to gynaecomastia (female characteristics). Hypersecretion of testosterone in women can lead to virilism, hirsutism and amenorrhoea (male characteristics)

Functions of the endocrine system

The endocrine system produces and secretes hormones in response to the changing needs of the body. It is responsible for: homeostasis, growth and sexual development.

Homeostasis

Homeostasis is the maintenance of a constant state. Examples of conditions within the body that need to be kept stable include:

- Body temperature
- Blood sugar levels
- Blood pressure
- Fluid balance
- pH

The production and secretion of hormones by the endocrine system helps to maintain homeostasis by controlling these conditions with the help of the nervous system. Together, the two systems ensure that the conditions within the body are kept stable in order to maintain physical and mental well-being. Any instability will result in ill-health.

Growth

The release of growth hormones in the pituitary gland determines the rate of growth of an individual.

Natural phases of growth include:

- Rapid growth during the first year of life
- Slow steady growth during childhood
- Rapid increase in growth during puberty
- Growth ceases at the average age of 16–17.

Gonadotrophins produced in the pituitary gland control the sexual differences associated with growth as males tend to be heavier and larger than females.

Sexual development

The human body goes through different stages of sexual development during a lifetime. These include puberty, menstrual cycle, pregnancy and menopause in a female and puberty in a male.

Remember

The average age for boys to reach puberty is between 13–16 whilst girls develop earlier reaching puberty generally around the ages of 10–14.

Puberty marks the onset of sexual development. The ovaries in females and the testes in males are stimulated into activity by the production of the gonadotrophins FHS and LH in the anterior lobe of the pituitary gland.

Puberty prepares the body for one of its most vital functions, that of reproduction. It does this by stimulating the production of sperm in a male and ova (eggs) in a female. In addition to this, every month the uterus in a female is prepared for the fertilisation of an egg to take place and the development of a baby. This is known as the menstrual cycle.

The **menstrual cycle** lasts for approximately 28 days and is controlled by the hypothalamus which stimulates the production of FHS and LH in the anterior pituitary gland. The menstrual cycle goes through three distinct phases:

Fascinating Fact

Some women are physically aware that ovulation has taken place. They experience a dull ache on one side of the abdomen on around the fourteenth day of their cycle.

1. Menstrual phase lasting for approximately five days. If an ovum is not fertilised, the lining of the uterus is forced to break down and levels of LH and progesterone are low. The menstrual flow contains the unfertilised ovum and the broken down lining together with blood and mucus.

2. Proliferative phase lasting for approximately nine days. FSH promotes the production of oestrogen which stimulates the lining of the uterus. At the end of this phase a new ovum is released from a follicle in an ovary and passes along the fallopian tube to the uterus. This is known as ovulation.

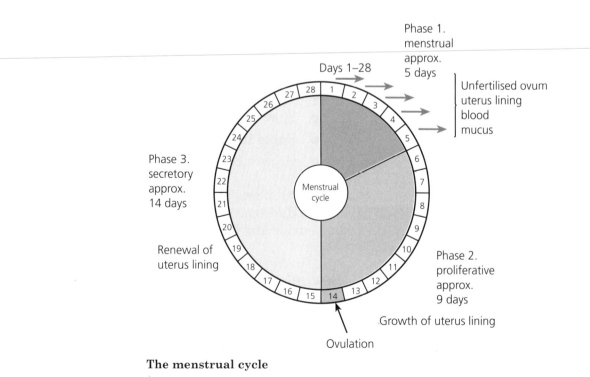

The menstrual cycle

Tip

The presence of HCG in the blood and urine forms the basis of a pregnancy test.

Fascinating Fact

The cry of a baby can be enough to stimulate the release of milk from the mother's breast!

Tip

The contraceptive pill prevents pregnancy in two ways. The mini pill contains progesterone which changes the uterus making it difficult for sperm to enter. The combined pill contains oestrogen and progesterone which stop ovulation.

Tip

HRT hormone replacement therapy is often prescribed to alleviate the symptoms of the menopause. Cycles of hormones are taken to correspond with the menstrual cycle.

3. Secretory phase lasting for about fourteen days. LH stimulates the renewal of the uterus lining which then produces progesterone stimulating the retention of fluid and the production of mucus. This prepares the area for entry by the sperm making it easier for them to reach the ovum for fertilisation to occur. If fertilisation does not occur, the whole cycle begins again.

Pregnancy occurs when fertilisation is successful and the ovum becomes embedded in the wall of the uterus and produces the hormone **HCG h**uman **c**horionic **g**onadotrophin. This maintains the lining of the uterus enabling it to continue producing progesterone which prevents menstruation and stops the production of new ova.

The placenta develops during the first 3–4 months and acts as a temporary endocrine gland producing hormones of its own as the pregnancy continues. The placenta forms a direct link between the circulation of the baby and its mother. During this time, the production of oestrogen and progesterone also stimulates the growth of the breasts. After the birth of the baby the anterior pituitary produces the hormone prolactin which stimulates the production of breast milk. This is forced out of the breasts by the stimulation of the hormone oxytocin from the posterior pituitary in response to the baby's needs.

Menopause marks the end of childbearing years in a woman and usually starts between the ages of 45–55; it can last for up to ten years, although the average is about five years. The breasts and ovaries decrease in size and become less responsive to the FSH and LH. As a result, hormone levels change and the menstrual cycle becomes irregular and eventually stops altogether. Other symptoms may include:

- Short-term flushing and sweating
- Loss of bone mass
- Loss of pubic and axillary hair
- Thinning of skin
- Mood swings.

Common conditions

An A–Z of common conditions affecting the nervous system

- ACROMEGALY – an over-secretion of the growth hormone GH in the anterior pituitary gland in adults which can result in a thickening of skin and an enlargement of the hands, feet and face, particularly the lower jaw.

- ADDISON'S DISEASE – disorder of the adrenal glands. There is an under-production of the hormones aldosterone and cortisol causing problems with metabolism and development. Muscle weakness, irregular menstrual cycle and dehydration are common symptoms.

- AMENORRHOEA – over-production of the male sex hormone testosterone in females caused by stress and dramatic weight loss, and resulting in loss of normal menstrual cycle.

- CHLOASMA – dark patches of skin which may appear as a result of a hypersecretion of MSH in the anterior pituitary gland.

- CONN'S SYNDROME – a condition which results from the over-secretion of aldosterone in the adrenal cortex and leading to high blood pressure due to high potassium levels in the blood. Kidney failure may also develop.

- CRANIAL DIABETES INSIPIDUS – an inability of the body to conserve water due to hyposecretion of ADH in the posterior pituitary gland.

- CRETINISM – see hypothyroidism.

- CUSHING'S SYNDROME – disorder of the adrenal glands. There is an over-production of the hormones aldosterone and cortisol causing the opposite effect from Addison's disease.

- DIABETES MELLITUS – there are two types of diabetes, type one and type two, both resulting in increased levels of sugar in the blood. Type one – insulin-dependent diabetes – occurs when there is an over-secretion of the hormone insulin produced in the islets of Langerhans in the pancreas. Type two – maturity onset diabetes – occurs when the tissues are unable to respond to the secretion of insulin. Type one develops at an early age and requires regular injections of insulin, while type two develops at a later age and can usually be controlled by changes in diet.

- DWARFISM – under-secretion of growth hormone GH from the anterior pituitary gland during childhood resulting in the failure of bones and organs to grow to normal size.
- ENDOMETRIOSIS – the development of cells from the lining of the uterus in an abnormal position. This may be caused by an imbalance in the production of the female sex hormones.
- FIBROIDS – the development of non-cancerous tumours in the wall of the uterus as a result of an imbalance in female sex hormones.
- GALACTORRHOEA – an excess of milk flow caused by an over-secretion of prolactin in the anterior pituitary gland.
- GIGANTISM – over-secretion of growth hormone GH from the anterior pituitary gland during childhood which results in an abnormal development in the length of long bones.
- GOITRE – enlargement of the thyroid gland caused by a lack of iodine.
- GRAVES DISEASE – see hyperthyroidism.
- GYNAECOMASTIA – over-production of oestrogen in males causing the development of breasts.
- HIRSUTISM – an over-production of testosterone in females leading to the development of male pattern hair growth. Common times for this include puberty, pregnancy and menopause.
- HYPERCALCAEMIA – excess calcium in the blood caused by the over-secretion of parathormone in the parathyroid glands.
- HYPERGLYCAEMIA – under-secretion of insulin and/or over-secretion of glucagons in the pancreas causing high blood sugar levels.
- HYPER PARATHYROIDISM – over-production of the hormone parathormone from the parathyroid glands which causes a decrease in calcium in the bones. This may lead to the condition osteoporosis whereby the bones become very brittle.
- HYPERTHYROIDISM – over-active thyroid production of the hormone thyroxine causing the metabolism to speed up and resulting in loss of weight, increased heart rate and swelling of the tissue behind the eyes. The eyes develop a prominent bulge. Hyperthyroidism is also known as GRAVES DISEASE and THYROTOXICOSIS.
- HYPOCALCAEMIA – lack of calcium in the blood caused by the under-secretion of parathormone in the parathyroid glands.

- HYPOGLYCAEMIA – an over-production of insulin causes low blood sugar levels with symptoms similar to drunkenness e.g. blurred vision, sweating, trembles and a lack of concentration. This condition may accompany diabetes.
- HYPO PARATHYROIDISM – under-production of the hormone parathormone from the parathyroid glands causing a decrease of calcium in the body.
- HYPOTHYROIDISM – under-active thyroid production of the hormone thyroxine causing the metabolism to slow down and resulting in weight gain. Hypothyroidism at birth results in the disorder CRETINISM and if untreated, leads to MYXOEDEMA in adults. The body systems work at a slower than normal pace.
- MYXOEDEMA – see hypothyroidism.
- POLYCYSTIC OVARIAN SYNDROME – under-production of LH in females resulting in the development of ovarian cysts. This is accompanied by an irregular menstrual cycle and sometimes infertility.
- PRE-MENSTRUAL SYNDROME (PMS) – mood swings, swollen, tender breasts, fluid retention and cravings for sweet foods accompany the lead up to menstruation in women.
- SAD (seasonal affective disorder) – an over-production of the hormone melatonin produced in the pineal gland. The result is depression and lethargy especially during the winter months.
- STRESS – prolonged negative stress results in long-term over production of the hormone adrenalin from the adrenal glands. This causes over-stimulation of some body systems e.g. muscles and under stimulation of other body systems e.g. digestion. The result is aching muscles and poor digestion.
- TETANY – abnormally low levels of calcium in the blood causing muscle spasms mainly in the hands, feet and face. May be caused by an under-production of parathormone.
- THYROTOXICOSIS – hyperthyroidism caused by the over-production of thyroxine.
- VIRILISM – an over-production of testosterone in females causing the onset of male characteristics including receding hair line, increased body and facial hair and the deepening of the voice. The menstrual cycle is also affected.
- VITILIGO – white patches of skin that may be caused by an under-secretion of MSH in the anterior pituitary gland.

System sorter

THE ENDOCRINE SYSTEM

Skeletal

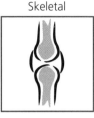

Muscular

Integumentary

The production of calcitonin and parathormone in the thyroid and parathyroid glands help to maintain calcium levels in the bones needed for strength and resilience.

Blood flow is increased in the muscles in response to the secretion of adrenalin from the adrenal glands which respond to stressful situations as part of the fight or flight syndrome.

Respiratory

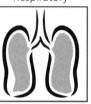

MSH melanocyte stimulating hormone produced in the anterior lobe of the pituitary gland stimulates the production of melanin in the stratum germinativum of the epidermis in response to the ultraviolet rays of the sun.

The effect of adrenal secretion increases breathing rates as more oxygen is needed in the body to fuel the muscles.

The nervous system is linked to the endocrine system by the hypothalamus and the pituitary gland. A pituitary stalk joins the two structures in the brain. The hypothalamus secretes hormones that activate the pituitary gland into stimulating the other endocrine glands.

Hormones are transported from the endocrine glands around the body in the blood. Hormones are suspended in the blood plasma.

Circulatory

The production of insulin and glycagon in the pancreas helps to regulate blood sugar levels. The sugar in the blood is processed by the digestive system.

The antidiuretic hormone vasopressin helps to regulate the fluid balance in the body by decreasing the amount of urine produced in the kidneys.

Nervous

Digestive

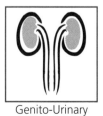

Genito-Urinary

The endocrine system provides the body with a centre for communication. Chemical messengers called hormones are produced by endocrine glands in different parts of the body. Hormones are passed directly into the blood stream to be transported to the part of the body they are to communicate with. Hormones are able to affect changes in other cells relating to the growth and development of the body.

Holistic harmony

The endocrine system needs to be very finely 'tuned' to maintain homeostasis, and many external physical factors as well as internal emotional factors contribute to the causes of imbalance. Care must therefore be taken to ensure that this system gets the balance it needs in the form of the following.

Angel advice

Limit the amount of alcohol and salted peanuts consumed at any one time. Alcohol is a diuretic and peanuts have a high salt content which contributes to the depletion of water in the body leading to problems of dehydration.

Fluid

Water intake is controlled by thirst. The hypothalamus is able to 'pick up' on the water levels of the body and stimulate the sensation of thirst when the levels become low. The production of ADH anti-diuretic hormone in the pituitary gland promotes the conservation of water in the body by decreasing the amount of urine produced in the kidneys.

Water retention and water excretion both affect blood volume which in turn affects blood pressure. Minerals like sodium play a part in fluid balance e.g. if sodium levels within the body are high, water is retained. The mineral ocorticoid aldosterone produced in the adrenal glands helps to maintain a balance of such minerals in the body.

Tip

The absorption of calcium by the digestive system is not very efficient but it is increased naturally during childhood when calcium is needed for the growth of bones and during pregnancy and lactation where calcium is needed for the development of the baby.

Nutrition

Calcium is an essential mineral needed by the body for a variety of functions including the formation of bones. Calcium is normally taken into the body in food and the excess is eliminated out of the body in urine. The hormones of the thyroid and parathyroid glands regulate this process with the secretion of calcitonin and parathormone. The normal recommended daily intake of calcium is about 1g and it is found in abundance within the foods that we eat especially in dairy products.

Vitamin D helps the body use calcium and is produced in the skin by sunlight (Chapter 2). Vitamin D is also found in oily fish.

Fascinating Fact

Intensively farmed soils can be deficient in iodine and may produce vegetables that are low in iodine for this reason.

Iodine is another essential mineral and is needed for the production of thyroxine in the thyroid gland in order to aid metabolism. Iodine is acquired from a person's diet; the recommended daily intake is 150mg. Iodine is found in meat and vegetables.

A balanced diet high in fresh, organic produce and low in animal fats, refined foods and caffeine will help to maintain hormone balance and prevent long-term health problems.

In contrast, a poor diet can aggravate hormone production resulting in imbalance e.g. the balance maintained between the female sex hormones oestrogen and progesterone is an important factor when considering the risks associated with breast cancer, endometriosis, ovarian cysts, fibroids and PMS (Chapter 8). High oestrogen levels can bring about feelings of depression, loss of libido and cravings for chocolate as well as the physical symptoms of fluid retention and breast swelling. In some cases change in diet may be all that is needed to alleviate the symptoms associated with hormone imbalance.

Angel advice

The use of meditation techniques at the start and/or end of treatments helps to focus the energy on a particular body system which, in a stressed client, will definitely include the endocrine system.

Rest

The benefits to the endocrine system of rest may be seen in the balance that the body experiences. Meditation is a particularly good method of resting both the body and the mind which will in turn have a calming effect on the endocrine system allowing it to function freely and effectively. The art of meditation involves switching off from the outside world and creating a safe distance from day-to-day problems for a short period of time. This state of rest soothes a stressed body and mind allowing vital time for a re-balancing effect to take place within the endocrine system.

Activity

Regular exercise ensures that the transportation of hormones is kept constant due to the stimulation of blood flow. In addition, the position of the spine in relation to the hips contributes to the well-being of the endocrine system. If posture is good then the endocrine glands are balanced and able to function more effectively. Yoga exercises including posture and breathing activities, are particularly useful in providing the body with the correct level of activity which is of direct benefit to the endocrine system. Activity can also strengthen the links between the endocrine and nervous systems. The human body as a whole benefits from the effective integration of all of its parts in its quest for homeostasis.

Air

Harmful chemicals can find their way into the body through the air that we breathe. Research has found that there are many synthetic chemicals that disrupt the endocrine system in this way. Some of the chemicals that make up the pollutants found in our environment today, e.g. some pesticides and herbicides, cause changes in metabolism, emotions and behaviour because of their effect on the endocrine system. These chemicals, known as xenoetrogens or 'oestrogen mimics', are also believed to be responsible for the reported drop in sperm count in men and the disruption in the balance of the female sex hormones. Care should be taken to ensure that the use of these chemicals is avoided or kept to a minimum for home use to prevent long-term problems.

Angel advice

The body is designed to deal with a certain amount of stress and this may be seen as positive stress, but it is wise to remember that the effects of negative stress can be threatening to our mental and physical survival.

Age

The ageing process is speeded up as a result of the over-secretion of some hormones. Stress initiates the secretion of adrenalin and noradrenalin from the adrenal glands. Whilst the actions of these hormones provide us with the 'instant energy' needed for 'fight or flight', their long-term effects have a definite downside. Functions such as digestion, repair and maintenance are slowed down in order to channel energy to other areas, e.g. muscles, and this has a 'knock on' effect on the well-being of the body and mind. As we age our natural responses to stress become slower and so the effects of stress become greater.

Colour

Activity

There are differing opinions as to whether or not the pineal gland is associated with the 6th or 7th chakra. The fact that the 6th chakra is associated with the third eye makes a strong case for the position of the pineal gland, which is also known as the third eye. Further reading will help you to make up your own mind.

The endocrine system is represented by a different colour for each of the following glands:

- Pituitary gland – violet
- Pineal gland – indigo
- Thyroid and parathyroid glands – blue
- Thymus gland – green
- Adrenal glands – red
- The pancreas is represented by yellow as part of the digestive system.
- The ovaries and testes are represented by orange as part of the reproductive system.

These colours can be used as a point of focus during treatment of the whole body in order to help the part e.g. an overworked adrenal gland will benefit from a full body massage because of the calming, relaxing effects massage has on the body as a whole. If at the same time the client is encouraged to visualise the colour red, not only will it take their mind away from their problems for a while, but it will also encourage the healing powers associated with colour on corresponding parts of the body.

In addition, each endocrine gland is strongly linked with a chakra of the corresponding colour.

- Pituitary gland – 7th or crown chakra
- Pineal gland – 6th or third eye chakra.
- Thyroid and parathyroid glands – 5th or throat chakra
- Thymus gland – 4th or heart chakra
- Pancreas – 3rd or solar plexus chakra
- Testes and ovaries – 2nd or sacral chakra
- Adrenal glands – 1st or base chakra.

Awareness

Identifying the sources of negative stress in our lives and forming strategies to make positive changes will help to maintain the balancing act that the endocrine system helps to control. The endocrine system, along with all other systems of the body, responds badly to constant overuse and misuse. Treatments can help to channel energy to the endocrine glands, which will in turn have a positive effect on the systems they control. It is important to take responsibility for the well-being of ourselves on a physical, emotional and spiritual level. Many clients are seeking treatments that focus on the harmony between body, mind and soul and regardless of our own beliefs we must respect the beliefs of others and be aware of the contribution our treatments make in fulfilling this need.

Special Care

A knowledge of the way in which the endocrine system works, together with an awareness of the factors which affect its general well-being, equip us with the tools to provide this vital system with the special care that it needs. The emotional trauma that accompanies the effects of hormonal imbalance causes distress in the lives of many people and could be alleviated if more attention was paid to the needs of the body and mind. No part of the body will ever be completely well unless the whole of the body is well and as the endocrine system has a controlling effect over the whole of the body, it makes sense to focus our special care in its favour. The rewards for such care will be experienced by body and mind allowing us the freedom to explore aspects of the human spirit that are as yet still unknown to us.

Treatment tracker

THE ENDOCRINE SYSTEM

Make up

Facials

Nail care

Changes in skin colour can occur due to hormone imbalance. The contraceptive pill can be associated with dark patches of skin known as chloasma. Make up can be used to successfully camouflage these pigmentation marks.

Facial and scalp massage helps to stimulate the circulation to the brain which in turn improves the links between the nervous and endocrine systems in the brain.

Hair removal

Over-production of the growth hormone in the thyroid gland results in the disorder acromegaly. Skin of the hands and feet thickens and nails become brittle and ridged. Nail care treatments can help to improve blood flow to the germinal matrix and improve surface texture

The use of electrical epilation is suitable for the treatment of hirsutism helping to remove the male pattern hair growth permanently in females.

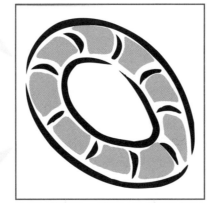

The 'stress busting' effect of massage helps to prevent excessive amounts of adrenalin secretion thus aiding physical and emotional well-being.

Electrical massage treatments like G5 and vacuum suction can help postural problems which in turn enables the endocrine system to work more effectively.

Working over the reflexes for each of the individual endocrine glands will help to regulate the functions of the body including – regulation of body rhythms, emotions, sexual desires and physical development.

Essential oils can be used to stimulate the endocrine glands to produce hormones to help to balance the body. Essential oils can also have a direct effect by the action of plant hormones on corresponding human hormones.

Massage

Electrical

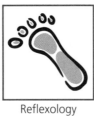

Reflexology

Aromatherapy

Beauty and holistic treatments improve all aspects of emotional, physical and spiritual well-being by helping to balance the endocrine system.

Knowledge review – The endocrine system

1 What is the difference between the nervous and the endocrine systems?

2 What is the name given to the structure that links the endocrine system with the nervous system?

3 What is a hormone?

4 What are hormones made from?

5 What is an exocrine gland and how does it differ from an endocrine gland?

6 Which endocrine gland is often referred to as the 'master gland' and why?

7 Name the hormones produced by the anterior lobe of the pituitary gland that have an effect on other endocrine glands.

8 Which endocrine gland is often referred to as the 'third eye' and why?

9 Which glands promote the actions of childbirth and the production of breast milk and where are they produced?

10 Why is the pancreas an exocrine and endocrine gland?

11 Which hormone contributes to the sunlight affective disorder SAD and where is it produced?

12 Which hormones contribute to the preparation of the body for 'fight or flight' and where are they produced?

13 Name the male and female sex hormones and the glands that produce them.

14 What are the three groups of steroids produced by the outer cortex of the adrenal gland called?

15 What is the thymus gland made of?

16 What are the actions of the hormones calcitonin and parathormone?

17 What happens to the metabolism as a result of an over- or under-active thyroid gland?

18 What may be prescribed to a female experiencing the symptoms of the menopause?

19 Which hormone helps to maintain fluid balance by conserving water in the body and decreasing the production of urine?

20 Name the three main functions of the endocrine system.

Glossary

absorption taking in a substance

acid mantle protective covering to the skin consisting of sweat, sebum and dead skin cells

acquired used to describe a disease that has been contracted after birth

actin thin protein myofilaments found in muscle fibres

active transport a means of passing a substance through the cell membrane

acute used to describe a disease that is sudden, severe and short in duration

adductors muscle group of the inner thigh

adenoids lymphatic tissue in the throat

adenosine triphosphate (ATP) energy produced in a muscle

adipose tissue fatty connective tissue found in the subcutaneous layer of skin

adrenal glands endocrine glands situated above the kidneys

adrenalin hormone produced by the adrenal glands. Associated with stress. 'Fight or flight' hormone produced by the inner medulla of the adrenal glands

adrenocorticotrophic hormone (ACTH) hormone produced in the anterior lobe of the pituitary gland

aerobic use of oxygen to create energy in muscles

afferent leading towards

agonist a muscle which is the prime mover

aldosterone hormone produced in the adrenal gland controlling levels of salts in the blood

alpha hydroxy acids (AHAs) fruit acids e.g. citric acid. Natural exfoliators

alveolar glands milk-producing glands in the breasts

alveoli tiny air sacs in the lungs where the interchange of gases takes place

ammonia waste product found in sweat

amylase digestive enzyme

anabolism the chemical reaction within a cell to form new parts e.g. protein

anaerobic production of energy in muscles without oxygen

anagen first stage of the hair growth cycle

anal canal final section of large intestine

anaphase part of the process of mitosis

anatomy structure

androgens male hormones

antagonist a muscle which relaxes whilst the prime mover or agonist contracts

anterior front of the body

antibody defence against an antigen

antidiuretic hormone (ADH) hormone produced in the posterior lobe of the pituitary gland to control water levels in the body

antigen harmful substance

anti oxidants nutrients that counteract free radical attack, i.e. vitamins A, C and E

antitoxin defence against a toxin

anus final part of large intestine

aorta main artery leading from the heart

apical breathing shallow breathing

apocrine glands sweat glands that open onto a hair follicle. Found in the groin, armpits, and breast areas and secrete the sweat associated with body odour

aponeurosis a broad, flat attachment of a muscle to a bone

appendix lymphoid tissue in the digestive system

arachnoid mater middle layer of the meninges in the CNS

areola pigmented skin surrounding the nipple

areolar tissue loose tissue found in the hypodermis providing strength, elasticity and support

arrector pili muscle tiny muscle attached to the hair follicle in the dermis. Contracts when body temperature decreases and pulls the skin into a goose pimple, lifting the hair and trapping warm air

arteries blood vessels leading away from the heart

arterioles small arteries

arthroscope thin tubular instrument used to examine a joint

atria upper chambers of the heart

auditory canal internal portion of the ear

autonomic nervous system (ANS) a combination of the sympathetic and parasympathetic nervous systems

axon nerve fibres which carry impulses away from the cell body

bacteria microbes that can cause illness and disease

ball and socket type of synovial joint allowing maximum movement between bones

basal layer (stratum germinativum) deepest layer of the epidermis

basement membrane point of attachment between two types of tissue

benign non-serious growth

bicep muscle of the front of the upper arm

bicuspid valve valve in the heart connecting the left atrium and ventricle

bile secretion of the liver which is stored in the gall bladder

bladder organ that stores urine

blood pressure pumping action of the heart

blood shunting changes in volume of blood in different parts of the body

blood vessels tubes of varying sizes transporting blood around the body. Smallest vessels are known as capillaries and largest are known as arteries and veins. Arterioles are smaller branches of arteries and venules are smaller branches of veins

bone dense connective tissue forming the skeletal system

bone marrow red bone marrow is found in cancellous bone and is responsible for the production of new blood cells. Yellow bone marrow is made up of fat cells and is stored in the length of some compact bone

Bowman's capsules cup-shaped part of a nephron

brachialis small muscle of the front of the arm

brain organ of the central nervous system

bronchi air passageways leading into the lungs

bronchioles small air passageways in the lungs

buccinator muscle of the face used when chewing or blowing

bursa a sac like structure found in some synovial joints to prevent excessive friction

caecum start of large intestine

calcitonin hormone produced in the thyroid gland

cancellous bone spongy bone

capillaries single cell structures at the end of blood and lymph vessels

carbohydrates energy producing food

carbon dioxide gas produced in the cells as a result of using oxygen

cardiac muscular tissue muscular tissue exclusive to the heart

cardiac sphincter ring of muscle between the oesophagus and the stomach

carotene yellow pigment affecting the colour of skin

carpals eight short bones of the wrist

cartilage connective tissue providing additional support for the skeletal system. HYALINE cartilage is found at the ends of bones at synovial joints. FIBROCARTILAGE is found in between bones at fibrous joints and ELASTIC CARTILAGE is found in the ear lobe

cartilaginous joints slightly moveable joints

catabolism chemical reactions within the cell that cause the break down of nutrients into the smallest possible form for energy production

catagen second stage of the hair growth cycle

cell microscopic part of an organism

cell membrane semi-permeable covering protecting a cell

central nervous system (CNS) the brain and spinal cord

centrioles a structure found within a cell associated with reproduction

cerebellum the small brain

cerebral hemispheres left and right halves of the brain

cerebrospinal fluid fluid that surrounds the brain and spinal cord

cerebrum the forebrain

cervical vertebrae seven bones of the spine forming the neck

cervix neck of the womb

chakra wheel of light/energy

chromosomes parts of a cell that contain the genetic make up of a person. There are 46 chromosomes in each cell

chronic used to describe a disease that is long in duration (the opposite of acute)

chyme broken down food in the stomach

cilia tiny hairs attached to a cell

ciliated cells cells containing cilia

circulatory systems controlling blood and lymph flow around the body in order to aid cellular metabolism

clavicle collar bone

clear layer (stratum lucidum) lying directly beneath the top layer of the epidermis

cocyx four fused bones of the spine forming the tail

collagen fibres connective tissue present in the dermis and responsible for giving the skin its youthful appearance. Formed from protein

colon large intestine

columnar cells tall cells forming simple epithelium

compact bone hard bone

compound epithelium stratified and transitional epithelium

concentric contraction shortening of muscle

condyloid type of synovial joint allowing flexion, extension, adduction, abduction and rotation movements

congenital a disease that is present from birth

connective tissue groups of cells which provide a protective and supportive function

contra indication a condition which prevents a treatment from taking place

Cooper's ligaments connective tissue supporting the female breasts

cornea transparent outer layer of the eye

corneum (stratum) top layer of the epidermis

coronary circulation blood circulation to and from the heart itself

corpus callosum connection between the two hemispheres of the cerebrum

corrugator supercilli muscle of the eyebrows

cortex middle layer of hair

cortisol mineralocorticoid produced in the adrenal glands

cortisone as above

coxae hip bones

cranial nerves twelve pairs of nerves coming from the brain to all parts of the face

cranium the head

cuboidal cells cube-shaped cells which form simple epithelium

cuticle, hair outer layer of the hair

cuticle, nail attaching nail fold to nail plate

cyanosis bluish colour to the skin as a result of decreased oxygen levels in the blood

cytokinesis part of the process of mitosis

cytoplasm semi fluid substance found in cells

deep under the surface

degenerative tending to deterioration

dehydration lack of moisture

deltoid muscle of the shoulder

dendrite nerve fibre which passes impulses to the cell body

deoxygenated lacking in oxygen i.e. blood

deoxyhaemoglobin haemoglobin that has picked up carbon dioxide from the cells

depilation temporary methods of hair removal

dermal cord connection between the papillary layer of the dermis and the hair follicle

dermal papilla part of the papillary layer of the dermis servicing the hair growth

dermis the true skin. Connective tissue lying directly below the epidermis

desquamate the shedding of epithelial skin cells

diaphragm muscle separating the abdomen and the thorax

diaphragmatic breathing deep breathing

diastolic pressure minimum pressure associated with the relaxation of the ventricles of the heart

diffusion the process of small molecules passing through the semi-permeable cell membrane

digestive system the body system associated with the intake and output of food and drink

dissolution the process of a fatty substance passing into the cell by dissolving into the cell membrane

distal the point furthest away from an attachment

diuretic a substance that increases urine production

DNA deoxyribonucleic acid

dorsi flexion bending the foot upwards

ducts tubes

duodenum first part of the small intestine

dura mater outer layer of the meninges of the CNS

eccentric contraction lengthening of muscle

eccrine glands sweat glands situated all over the body and responsible for assisting the body's temperature control

effector a muscle or an organ that will be stimulated to perform an action by motor nerves

efferent leading away

elastin fibres connective tissue present in the dermis providing the skin with elasticity

ellipsoid type of synovial joint allowing flexion, extension, adduction and abduction movements

embryo early stages of a fertilised ovum

emulsification oil suspended in water

endocardium inner layer of the heart

endocrine glands glands lined with epithelial tissue which secrete a substance directly into the blood

endocrine system controlling the body through chemical messages (hormones)

endomysium connective tissue surrounding muscle fibres

endoplasmic reticulum channels transporting substances within a cell

endorphins neuro transmitters associated with raising the pain threshold

enzyme a protein produced in a cell that is capable of speeding up a chemical reaction for which it is responsible

epidermis epithelial tissue making up the top layer of skin. Consists of five separate layers

epididymis coiled tube leading from the testes to the vas deferens

epiglottis cartilage forming a lid-like structure in the throat

epilation permanent method of hair removal

epimysium connective tissue surrounding a complete muscle

epithelial tissue groups of cells providing a protective function

eponychium living part of the cuticle of the nail

erepsin digestive enzyme

ergosterol fatty substance found in the skin which is converted to vitamin D in response to stimulation from ultra violet rays

errector spinae group of muscles which extend along the spine

erythema dilation of surface capillaries through stimulation causing the skin to redden

erythrocytes red blood cells

ethmoid bone of the nasal cavity

eustachian tube tube connecting the nose to the ear

eversion turning the foot out

excretion the process of eliminating waste from the body

exfoliators skin care products containing granules which gently remove excess dead skin cells when worked into the skin

exocrine glands glands lined with epithelial tissue that secrete substances through a duct to another part of the body

expiration breathing out

extensor a muscle causing extension (straightening)

flexor a muscle causing flexion (bending)

external on the outside of the body

extracellular fluid body fluids e.g. blood, mucus

faeces waste product of the digestive system

fallopian tubes passageway leading from the ovaries to the uterus

fascia layer

fast twitch fibres muscle fibres that produce rapid contractions

fats foods that provide the body with fuel

femur long bone of the upper leg

fertilisation the impregnation of an ovum by a sperm

fibre food which is indigestible and is needed to aid elimination of waste from the digestive system

fibroblast cells cells responsible for forming new fibrous tissue in the dermis in the event of injury

fibrous joints fixed joints

fibrous tissue tough connective tissue

fibula one of the bones of the lower leg

filtration a way of moving fluid through the cell membrane using pressure

fixator a group of muscles which fix the position of the body

flexor a muscle causing flexion (bending)

foetus developing baby

follicle (hair) a small pocket situated in the dermis in which a hair develops and grows

follicle stimulating hormone (FSH) hormone produced in the anterior lobe of the pituitary gland

free edge tip of the nail that extends over the region of the nail bed

free radicals the toxic by-product of energy metabolism which contributes to the premature ageing of the skin

frontal bone of the forehead

fungi a microbe that causes illness and disease

gall bladder accessory organ of the digestive system. Stores bile to aid in digestion

gastrocnemius large muscle of the calf

genetic inherited

genitalia reproductive organs

germinal matrix living, reproducing part of the hair and nails

germinativum (stratum) the deepest layer of the epidermis where cellular regeneration takes place

gland a structure lined with epithelial tissue which secretes a substance

glomerulus tight knot of capillaries found in the kidneys

glucagon hormone produced in the pancreas

gluteals muscle group of the buttocks

glycogen the form in which carbohydrates are stored in the muscles and the liver

goblet cells cells found in simple epithelium, which produce mucus

golgi body part of a cell that processes protein

gonadotrophins hormones produced in the anterior lobe of the pituitary gland which have a stimulating effect on the gonads

gonads reproductive organ

goose bump tiny bump on the surface of the skin which arises as a result of the arrector pili muscle contracting

granular layer (stratum granulosum) The layer of the epidermis lying directly above the stratum germinativum. Keratinisation begins in this layer

granulosum (stratum) as above

growth hormone (GH) hormone produced in the anterior lobe of the pituitary gland

haemoglobin substance that allows the blood cells to carry oxygen and carbon dioxide

hair bulb base of the follicle where a new hair develops

hamstrings muscle group of the back of the thigh

haversian systems part of the formation of bone

helix upper section of the external ear

hepatic flexure bend in the large intestine (liver side of the body)

hilum inner section of the kidneys

hinge type of synovial joint allowing flexion and extension movements

histamine present in all body tissue and released in the skin in response to sensitivity

holistic considering the complete person – the whole self – body, mind and soul

homeostasis physiological stability

hormones chemical messengers produced by the endocrine glands

horny layer (stratum corneum) The top layer of the epidermis consisting of dead, flat, hard cells

human chorionic gonadotrophin (HCG) hormone associated with pregnancy

humerus bone of the upper arm

hyaline cartilage tough fibrous connective tissue

hydrated absorbed of moisture

hyoid bone to which the tongue attaches

hypersecretion over-secretion

hypodermis the fatty layer also known as the subcutaneous or subcutis layer

hyponychium area of cuticle lying underneath the free edge of the nail

hyposecretion under-secretion

hypothalamus part of the cerebrum that links the nervous and endocrine systems together

ileocaecal sphincter ring of muscle at the end of the ileum

ileum final part of the small intestine

ilium flat bone of the coxa

immune system the body system responsible for protecting the body against disease

infectious refers to a disease that can be passed from person to person

inferior at the lower end of the body

infestation worms, insects and mites that attack the body

inflammation redness, heat, pain, swelling and loss of function of a part of the body

insertion the movable point of attachment of muscle to a bone

inspiration breathing in

insulin hormone produced in the pancreas

integumentary system the skin, hair and nails

intercostals muscles lying between the ribs

internal inside of the body

interphase a part of the process of mitosis

interstitial fluid fluid that bathes the cells

intracellular fluid fluid found within the cells

inversion turning the foot in

involuntary muscular tissue smooth muscular tissue responsible for involuntary action e.g. peristalsis

iridology the physical study of the eye to determine a person's state of health

iris coloured part of the eye

ischium flat bone of the coxa

islets of Langerhans endocrine section of the pancreas

isometric static muscle contraction without movement at a joint

isotonic active muscle contraction with movement at a joint

jejunum middle section of the small intestine

joint the point at which two or more bones meet

keratin protein found in the skin, hair and nails

keratinisation the development of keratin

keratinocytes cells that produce keratin

kidneys two bean shaped organs that produce urine

lacrimal bones forming the eye sockets

lactase digestive enzyme

lacteals lymphatic capillaries in the small intestine

lactic acid the waste product of energy production in a muscle

lactiferous sinuses storage space for milk in the female breasts

lamellae thin plates of bone tissue

lanugo hair hair that is present on the body prior to birth

larynx upper throat

lateral away from the midline

lateral costal breathing normal breathing

lateral longitudinal arch arch running along the outside of the foot

latissimus dorsi muscle of the back

leucocytes white blood cells

ligaments connective tissue holding bones at a joint

limbic centre part of the cerebrum connected with the sense of smell

lipase digestive enzyme

liver largest internal organ

local refers to a disease that is limited to one area of the body

longitudinal lengthways

loop of Henle a loop formed in a nephron in the kidneys

lucidum (stratum) clear layer of the epidermis

lumbar vertebrae five bones of the spine forming the lower back

lumen central cavity of blood vessels

lunula half moon area of living cells at the base of the nail

lungs organs of respiration

luteinising hormone (LH) hormone produced in the anterior lobe of the pituitary gland

lymphatic tissue connective tissue which forms lymph nodes

lymphocytes white blood cells that produce an antibody against an antigen

lysosomes the 'disposal unit' of a cell

macrophages cells that destroy antigens

malignant tendency to become progressively worse resulting in death

maltase digestive enzyme

mammary glands female breasts

mandible lower jaw bone

masseter muscle located in the cheek which aids the action of chewing

mast cells produce histamine when skin is damaged or irritated

master gland pituitary gland

mastication chewing

maxillae bones of the upper jaw

medial towards the midline

medial longitudinal arch arch running along the inside of the foot

medulla inner most layer of the hair. Not always present

medulla oblongata forms the lower part of the brain stem

meiosis a process of cell reproduction to create a new organism

melanin the natural colour pigment

melanocyte stimulating hormone (MSH) hormone produced in the anterior lobe of the pituitary gland

melanocytes cells which produce melanin

melatonin hormone produced in the pineal gland

meninges protective layers of the CNS

menopause the time when the female menstrual cycle comes to an end

menstrual cycle monthly release of an ovum from the ovaries

menstruation the release of an unfertilised ovum from the uterus

mentalis muscle at the top of the chin

metabolism chemical process within cells

metacarpals five bones of the hand

metaphase a part of the process of mitosis

metatarsals five bones of the feet

microbes a minute disease-causing organism

micturition the passing of urine

mid brain part of the central nervous system linking the brain and the spinal cord

midline the centre line of the body from head to toe

mineralocorticoids hormones produced in the cortex of the adrenal glands

minerals food that provides the body with the nutrients needed to maintain its functions

mitochondria muscle 'power houses' storing glycogen and oxygen

mitosis simple cell division

moisturiser skin care product used to restore fluid balance and to provide protection to the surface of the skin

molecule simplest freely existing chemical unit

monocytes white blood cells

motor end plate point of a nerve ending attached to a muscle fibre

motor nerves receive messages from the central nervous system

motor point point where a motor nerve enters a muscle

mucosa layer inner most layer of the alimentary canal

mucus fluid formed by goblet cells

muscle fatigue muscles that are over-worked

muscle tone partial contraction of muscles

muscular system responsible for movement

muscular tissue cardiac, visceral and skeletal tissue

myelin fatty substance forming a sheath around a neuron

myoblasts muscle forming cells

myocardium middle layer of the heart

myofibrils threadlike structures that form muscles

myofilaments protein filaments (actin and myosin) that make up muscles

myoglobin muscle store of oxygen

myosin thick protein myofilaments found in muscle fibres

nail bed skin lying directly below the nail

nail fold base of the nail where the new cells are produced

nail grooves tracks found either side of the nail guiding its growth

nail plate hard, clear structure forming the nails of the fingers and toes

nail wall skin which curls over the nail grooves at the side of the nail

nasal bones of the bridge of the nose

nasalis muscle covering the front of the nose

nephron small tubes found in the kidneys

nervous system this system helps to control the body by sending electrical messages (impulses)

nervous tissue groups of neurons and neuroglia forming the nervous system

neuroglia cells which support the neurons

neuron nerve cell

neurotransmitter chemical released at a nerve synapse

node of Ranvier the junction in the myelin sheath of the axon of a neurone

nipple external portion of the breast

noradrenaline hormone produced by the inner medulla of the adrenal glands

nucleoplasm a form of protoplasm that surrounds the nucleus of a cell

nucleus the centre of a cell that contains the information necessary for cell development

obliques muscle group of the waist

occipital bone of the back of the head

occipito-frontalis muscles of the scalp

oedema swelling caused by excess fluid in the tissues

oesophagus tube which carries food from the throat to the stomach

oestrogen female hormone produced in the ovaries

olfactory cells, bulb, tract and **nerves** associated with the nose and the sense of smell

orbicularis occuli circular muscle of the eye

orbicularis oris circular muscle of the mouth

organ a structure that is made up of two or more tissue types and has specific form and function

organelle little organs found within cells

organism the sum of cells, tissues, organs and body systems

origin fixed point of attachment of a muscle to a bone

osmosis a way of moving diluted substances through the cell membrane

ossification the process by which bones are developed

osteoblasts bone-forming cells

osteoclasts special cells that break down old bone cells

osteocytes mature bone cells

ova eggs

ovaries female gonads

ovum single egg

oxidation to combine with oxygen or to remove hydrogen

oxygen vital gas breathed into the body from the atmosphere ('life force')

oxygenated rich in oxygen i.e. blood

oxyhaemoglobin haemoglobin that has been oxygenated

oxytocin (OT) hormone produced in the posterior lobe of the pituitary gland

pace maker artificial structure used to regulate the heartbeat

palate roof of the mouth

palatine bones of the roof of the mouth

pancreas accessory organ of digestion. Also has endocrine functions

papillary layer the top layer of the dermis that connects with the lowest layer of the epidermis

parasites animal or vegetable organism which lives on or within another

parathormone (PTH) hormone produced in the parathyroid glands

parathyroid glands endocrine glands located in the neck along with the thyroid gland

parietal bones of the crown of the head

patella sesemoid bone of the kneecap

pathogenic disease producing

pectorals muscles of the upper chest (major and minor)

penis male sexual organ

pepsin digestive enzyme

pericardium outer layer of the heart

perimysium connective tissue surrounding bundles of muscle fibres

perionychium cuticle which extends up the sides of the nails

periostium connective tissue covering and protecting bones

peripheral nervous system (PNS) Twelve pairs of cranial nerves and 31 pairs of spinal nerves

peristalsis involuntary muscular action

peritoneum outer layer of the alimentary canal

Peyer's patches lymphatic tissue in the small intestine

phagocytic cells cells capable of destroying bacteria and foreign matter

phalanges bones of the fingers, thumbs and toes

pharynx back of the throat

photosynthesis plant respiration

physiology functions

pia mater inner layer of the meninges of the CNS

pineal gland endocrine gland situated in the brain

pinna earlobe

pituitary gland endocrine gland situated in the brain. Also known as the master gland

pivot type of synovial joint allowing rotation movement

plane type of synovial joint allowing gliding movements only

plantar flexion stretching the foot downwards

plasma the fluid part of blood

platysma muscle of the neck

pleura outer protective covering of the lungs

plexus network of nerves

pons varolii forms a bridge between the two hemispheres of the brain

pore opening onto the skin

portal circulation blood flow from the digestive system to the liver

prickle cell layer (stratum spinosum) a layer of the epidermis

prime mover the muscle under contraction (agonist)

procerus muscle covering the bridge of the nose

progesterone female hormone produced in the ovaries

prolactin (PRL) hormone produced in the anterior lobe of the pituitary gland

pronate turn the hand so that the palm is facing downwards

prone lying face down

prophase a part of the process of mitosis

prostate gland produces the fluid part of semen in a male

proteins food that provides the body with the nutrients for growth and repair

protoplasm a transparent, jelly-like substance from which cells are made

proximal closest point to an attachment

puberty the time when secondary sexual characteristics start to develop

pubis flat bone of the coxae

pulmonary circulation blood flow to and from the lungs

pulse contraction and relaxation of the ventricles of the heart

pupil centre of the iris of the eye

pyloric sphincter ring of muscle between the stomach and duodenum

pyruvic acid waste product produced in muscles

quadriceps muscle group of the front of the thigh

radius one of the bones of the forearm

rectum latter part of large intestine

rectus abdominus muscle of the abdomen

reflex quick reaction

renal pelvis part of the kidney

renin enzyme produced in the kidney to regulate blood pressure

respiration the use of oxygen and the release of carbon dioxide

respiratory system the body system responsible for breathing in and breathing out air

reticular layer the second layer of the dermis

reticulin fibres fibres found in the reticular layer of the dermis responsible for providing support to the structures found within this layer

retina the inner most layer of the eye

ribosomes the 'protein houses' of a cell

ribs twelve pairs of bones forming part of the thorax

right lymphatic duct small lymph duct of the body

risorius muscle of the lower cheek responsible for smiling and grinning

RSI repetitive strain injury

rugae folds in the mucosa lining of the stomach

sacrum five fused vertebrae of the spine forming the base of the spine

saddle type of synovial joint allowing flexion, extension, adduction, abduction and rotational movements

safe stress high demands, low constraints with high levels of support

saliva fluid produced by salivary glands

salivary glands produce the fluid saliva to aid digestion of food

sartorius one of the muscles of the front of the thigh

scapulae shoulder blades

Schwann cell forms the myelin sheath of a neurone

scrotum muscular sac holding the testes

sebaceous gland small gland opening onto a hair follicle in the dermis and responsible for secreting sebum

sebum the skin's natural oil

secretion a cellular process for releasing a substance

semen secretion from the testes, seminal vesicles and prostate gland containing sperm

seminal vesicles responsible for secreting the fluid part of semen

sensory nerves link with the central nervous system

septum a division or partition

serratus anterior muscle located under the arm

sesemoid small bones located within tendons e.g. patella, hyoid bones

sex corticoids hormones produced in cortex of the adrenal glands

simple epithelium single layer of cells

sinus air space in bones containing mucous lining

skeletal muscle muscles associated with movement of the body

skeletal system responsible for body shape and the attachment of muscles

skull bones of the cranium and the face

slow twitch fibres muscle fibres which produce slow contractions

smooth muscular tissue muscles associated with the movement of internal organs

solar plexus network of nerves situated in the abdomen below the diaphragm

soleus small muscle of the calf

spermatozoa sperm

sphenoid bone of the eye socket

sphincter muscle circular muscle surrounding an opening

sphygmomanometer a device to monitor blood pressure

spinal nerves thirty-one pairs of nerves running from the spine to all parts of the body

spine bones of the neck and back

spinosum (stratum) prickle cell layer of the epidermis

spleen lymphatic tissue situated in the upper left side of the abdomen

splenic flexure bend in the large intestine (spleen side of body)

squamous cells flattened cells that form simple epithelial tissue

sterno cleido mastoid muscles of the neck

sternum breast plate

stratified formed in layers

stratum layer

stress high demands, high constraints with low levels of support

striated having stripes e.g. skeletal muscular tissue

subarachnoid space middle layer of the meninges of the CNS

subcutaneous layer fatty layer

subcutis layer another name for subcutaneous layer

submucosa second layer of the alimentary canal lying beneath the mucosa layer

sucrose digestive enzyme

sudoriferous glands sweat glands

superficial close to the surface

superior above

supinate turn the hand so that the palm is facing upwards

supine laying face up

sutures joins in bones

synapse junction between neurons

synergist small muscles which help movement

synovial joint freely movable joint

systemic used to describe a disease that affects the whole body

systemic circulation blood flow between the heart and the whole body

systolic pressure maximum pressure associated with the contraction of the ventricles of the heart

tarsals seven bones that make up the ankle

taste buds nerve endings in the tongue associated with the sense of taste

telogen final stage of the hair growth cycle

telophase a part of the process of mitosis

temporal bones of the temple

temporalis muscle of the temple region of the face

tendons connective tissue attaching muscles to bones

terminal hair thick, coarse hair e.g. head hair

testes male gonads

testosterone male hormone produced in the testes

thoracic duct main lymph duct of the body

thoracic vertebrae 12 bones of the spine forming the upper and mid back

thorax bones of the chest

thrombocytes platelets

thymosins hormones produced by the thymus gland

thymus gland endocrine gland in the thorax associated with the immune system

thyroid gland endocrine gland situated in the neck

thyroid stimulating hormone (TSH) hormone produced in the anterior lobe of the pituitary gland

thyroxine hormone produced in the thyroid gland

tibia one of the bones of the lower leg

tibialis anterior muscle forming the shin

tissue groups of cells with the same function

toner skin care product used after a cleanser to help balance the surface of the skin

tonsils lymphatic tissue in the throat

trachea windpipe

transitional cells compound epithelium that is expandable

transverse running across

transverse arch arch running across the foot

trapezius large triangular muscle of the back

triangularis muscle drawing the angles of the mouth downwards

tricep muscle of the back of the upper arm

tricuspid valve valve in the heart connecting the right atrium and ventricle

triiodothyronine hormone produced in the thyroid gland

trypsin digestive enzyme

tumour a growth resulting from an over-production of cells

turbinate bones forming the sides of the nose

ulna one of the bones of the forearm

urea waste product found in sweat

ureters tubes leading from the kidneys to the bladder

urethra tube leading from the bladder out of the body

uric acid waste product found in sweat

urine water and waste produced in the kidneys and released from the bladder

uterus womb

uvula arch of the soft palate in the mouth

vagina passageway leading from the uterus to the outside of the female body

vas deferens tube leading from the testes to the urethra in a male

vaso constriction tightening of blood capillaries

vaso dilation widening of blood capillaries

veins blood vessels transporting blood to the heart

vellus hair the downy hair that covers most of the body

ventricles lower chambers of the heart

venules small veins

vertebrae bones of the spine

villi tiny projections in the mucosa layer of the alimentary canal

virus a microbe that is responsible for illness and disease

visceral muscular tissue smooth muscular tissue providing involuntary movement

viscous thick, sticky fluid

vitamins food that provides the body with nutrients to maintain its vital functions

voluntary muscular tissue skeletal muscle

vomer bone forming the top of the nose

womb uterus

zygomatic bones of the cheeks

zygomaticus muscle of the cheek

zygote a complete cell that has been formed by the fusing together of a female egg and a male sperm

Index